Henry Frith

The Romance of Engineering

Stories of the highway, the waterway, the railway, and the subway

Henry Frith

The Romance of Engineering
Stories of the highway, the waterway, the railway, and the subway

ISBN/EAN: 9783337344764

Printed in Europe, USA, Canada, Australia, Japan

Cover: Foto ©berggeist007 / pixelio.de

More available books at **www.hansebooks.com**

THE

ROMANCE OF ENGINEERING;

STORIES OF

THE HIGHWAY, THE WATERWAY, THE RAILWAY, AND THE SUBWAY.

BY

HENRY FRITH,

Author of "The Biography of a Locomotive," "The Triumphs of Modern Engineering," "On the Wings of the Wind," "The Search for the Talisman," etc., etc.

WITH ABOUT ONE HUNDRED AND FIFTY ILLUSTRATIONS.

LONDON:

WARD, LOCK & CO., LIMITED,

WARWICK HOUSE, SALISBURY SQUARE, E.C.

NEW YORK AND MELBOURNE.

PREFACE.

IT was inevitable that the writer of this volume, while passing over some old ground, in his search for the Sources of Information, should have occasionally struck into the channels formed by his predecessors, while projecting a line of his own. But the Author can conscientiously declare that he has at all times penetrated to the fountain-head for his facts, and has not tapped the reservoirs of others on his way. In any instances in which the opinions or views of writers are quoted, due acknowledgment is made in the text or by notes.

This volume, as its title indicates, is a Romance of Engineering; a consideration of the non-technical and anecdotal aspect of road-making, railway-making, and other early Engineering Works. It is not intended, primarily, for students, but for the large classes of youthful and middle-aged readers who desire to know something of the means in the past which have tended to secure the civilization of the present.

H. F.

CONTENTS.

—◆◆—

THE ROMANCE OF THE HIGHWAY.

Contents.

THE ROMANCE OF THE RAILWAY.

Contents.

THE ROMANCE OF THE SUBWAY.

LIST OF FULL PAGE PLATES.

LIST OF ILLUSTRATIONS IN THE TEXT.

EARLY STONE BRIDGE IN ENGLAND, AT STRATFORD-LE-BOW.

TIIE ROMANCE OF THE HIGHWAY.

CHAPTER I.

ANCIENT ROADS AND PATHS. — OLD BRIDGES AND THEIR
ASSOCIATIONS.—FORDS AND FERRIES.

TO the nineteenth century, now passing to its rest, must be accorded the palm for invention, and for the stupendous nature of engineering works. Look in whatever direction we may, the magnificent monuments of the engineer meet our gaze, whether on land or on the sea, in motion or at rest. These structures are the practical outcome of the skill and scientific research of clever men: the movements and the continued existence of other structures—ships, locomotives, machinery—are dependent upon the knowledge of other clever men in a lower grade—men of trained senses and keen insight—mechanics,

1   B

to whom we owe the development of nearly all the greatest discoveries in steam, and in its application. From the time of Savery to the present day the "working man" has been the foremost in the race of invention, of mechanical powers and their application. The obviously practical side of all these wondrous works — whether roads, bridges, viaducts, canals, railroads, steam engines, and other machines, lighthouses, and so on — has been frequently presented to us, and we have been interested in reading of those triumphs of the engineer.

It is now our turn to endeavour to put before readers the romantic side of these constructions; to look back on the olden time, and trace in the spirit of incident and adventure the progress and final success of the engineer and his creations on the highway, the waterway, the railway, and the sub-way. We might extend our researches even into the atmosphere and the "air-way," and narrate the escapes, dangers, and adventurous achievements of those who have voyaged and prospected in the air, as the engineer has on and in the earth.

Although this is an exceedingly practical age, although business banishes sentiment, and romance is designated nonsense, the Romance of Engineering is continually to be seen in our most commonplace work and every-day toil. Take only the railroad as an example. What adventure and romance is connected with our dull, but swift, trains, and with the dead level of the permanent way ! So, to those who seek them, Romance and Adventure will be found amid these unpromising surroundings, and will reward the searcher who will carefully sift the material amid which they lie hidden.

This we propose to do, and now let us strike without further preface into the history of our highways in which so many romantic incidents may be found.

"Those inventions which abridge distances have done much for the civilisation of our species," wrote Lord Macaulay ; and this is an evident conclusion. Roads of some kind were made in Egypt, but not by the will of the people. Like the Simplon road, these ancient highways, whether constructed by the

Egyptians, or the Incas of Peru, were constructed generally for military or regal purposes by the sovereign or despot of the country, for the purposes of conquest or his own aggrandisement.[1]

These so-called roads were generally only tracks, and the " king's highway " of the Bible only a path. To the Carthaginians is attributed the science of road-making, and from them the Romans, and perhaps the Britons, learned the art. These Roman roads are still traceable, the word "street" indicating their paved route. The Prætorian or military roads were, however, distinct from the Consular or public roads; and so strictly were these distinctions observed and maintained, that the roads sometimes ran parallel to each other, as rival railway lines do in the present day.

But if the uses of the roads differed, the manner of construction also differed. The military road was wide, and elevated in the centre, partly paved; with blocks of stone at frequent intervals, to facilitate the mounting of the cavalry. The materials for engineering these Prætorian roads were, of course, drawn from the neighbourhoods through which the road was run, and most solid was the foundation.

The Consular, or public road, was broad, and direct in course. The paved centre was raised, the "causeway" or *chaussé* some ten feet wide. The maintenance of these roads fell upon the landowners who gave their names to the sections, notwithstanding that the whole thoroughfare was named after the consul who built it. The country roads crossed the others at right angles, and gate-houses were erected at the crossways. The solidity of the Roman roads is well known, and their direction from and to the capital is expressed in the proverb, " All roads lead to Rome."

The pavements were also distinguished : the simple paved roads (*stratæ viæ*) of gravel and pebble ; the flint pavements of

---

[1] We read that in Peru, in ancient times, runners proceeded in relays over the wonderful highways of the Andes, from 150 to 200 miles a day, to carry fresh fish to the royal table.

--

the *viæ silice stratæ* ; and the *viæ saxo et lapide quadrate stratæ*, or streets, which were, as indicated, paved with square, flat stones, perhaps united by Roman cement.

The care taken in the construction of these ways was remarkable. The centre was paved, but on either side of the pavement there was a soft sandy or clay track for the convenience of the beasts of burden, and with a certain convexity to insure the rapid draining off of rain-water.

Older roads are found in England—the Fosse-Way, in Somersetshire, for instance—and it would appear that great care was always taken by the Romans to raise the highways so as to preserve them from flood. There are several of these very ancient roads or causeways in England—roads probably originally made by the Britons and improved by the Romans. The principal are the Watling Street, the Fosse-Way, Ermin Street, and Ikenild Street, all leading out from London, from the place where rests the London Stone, in Cannon Street, in various directions.

Watling Street, " *Via Vitelliam,*" as the Romans called it, started northwards to the site of St. Paul's Churchyard,[1] and was continued by the Romans, who of course adopted or used such portions of the existing roads as they deemed necessary. The great distinction in the engineering of the British and Roman road is the direction. The former always kept high and adapted itself to the country ; the latter went straight across the land.

Ermin Street led to Colchester, and thence to Chester ; and Ikenild, or the "old street to the Iceni territory," ran into Essex. It was the Essex Road, and may be found preserved in Old Street still.

When the Romans quitted Britain, their highways remained, but by degrees got choked by brushwood and seedling. Time and the " elements " militated against these solid constructions.

---

[1] Antiquarians say that Watling Street was the street to the wattled or hurdled enclosure in which stood the temple of Diana, where St. Paul's now stands.

Wind and rain, storm and tempest, strove against them. The breeze brought the seed, and grass grew; the rain washed off the sloping earth-sides. The undermined blocks of granite sank reluctantly down, and the weeds grew apace over these gravestones of Roman progress. So the streets sank out of sight, and only by accident were they discovered.

Geoffrey of Monmouth writes of Belinus, who built the wondrous gate whose ancient site is commemorated by the fish-market, that he summoned all the workmen of Britain and commanded them to pave a causeway of stone and mortar which should run the whole length of the island from the sea of Cornwall to the shore of Caithness, passing the cities *en route*. Another he ordered to cross England, and other two in more slanting directions.

But after the departure of the Romans, road-making as a practice disappeared. Continuous traffic in some places had worn the Ridge-ways into Hollow-ways; so that the once high track ran between banks, as in a cutting. These tracks were not paved roads, and consequently disintegrated rapidly. The pilgrim, the minstrel, the knight, and the friar plodded or rode along these soft and springy paths; cavalcades of men and women rode, *en cavalier*, too, over these bosky tracts between the hedges, or clattered upon the stony causeway to the ford.

There were few, if any, bridges then, as our town nomenclature can testify — Stratford and Stony-Stratford with many others confirm this. When the body of the sainted Bishop Erkenwald was being conveyed from the abbey of Barking to London, the *cortége* could not cross the Lea without danger, and a miracle. Stowe relates a fatal accident which befell the attendants of the queen of Henry the First at the Lea passage, known as the Old Ford, where Maud herself was "well washed in the water!" This accident led to the building of two stone bridges "of the which one was situated over Lue, at the head of the town of Stratford, now called Bow, because the bridge was arched like a bow, a rare piece of work; before that time the like had never been seen in England!"

The existence of these fords accounts for the number of towns whose names end in "ford." They were built by the river passage, the meeting place of the travellers, where merchandise was exchanged and news communicated. The roads were then unsafe ; people travelled in company, as the ways were so overhung by trees and strewed with brushwood that it was a matter of some difficulty and danger to traverse them. Young readers of "Ivanhoe" have no doubt remarked upon the ease with which outlaws and foresters used to conceal themselves on the public ways, but the "cover" was close and thick, robbery was easy, being frequently connived at by the lord of the manor, or by the baron in possession.

When the claim of the traveller to protection was at length recognised in the thirteenth century, we find a statute was ordained by which the bushes alongside the road were commanded to be cut down to prevent robbers from concealing themselves. This was the first indication of the tendency of the Englishman to mend his ways, and to turn in the direction of road-engineering, after the departure of the Romans.

From what has been said, it may be gathered that the Romans did not cut the first roads in England, though they "made" them. The Ridge-ways and the Hollow-ways, the upper and the out-trodden tracks, still indicate the old paths by which the people got about before and after the departure of the Romans. But that the ancient Britons cut roads cannot be gainsaid, for they had chariots and other wheeled vehicles, which ran on the turfy tracks and so-called "streets."

But in numerous instances these tracks and fords were useless. We have seen how Maud was treated at Bow, and before that time how the bishop's body was only taken over the Lea "by a miracle." The old stepping-stones were often submerged, and the rude planks or trunks of trees were so frequently washed away, that something of a more permanent character was found necessary.

I suppose our first impression of a bridge was derived, in childhood, from the old willow-pattern plate. In the immortal scene depicted upon it, conveying the old love story, the bridge is a conspicuous feature. This faithfully pourtrays the Chinese bridge of steep approach, as most early bridges were. The form, the arch, was perhaps adopted from the rainbow, that spans the cloud with brilliant hues.

But whatever its origin, the span of progress from the stepping-stone, the tree, or the suspended twigs, to the Forth or Niagara Bridges of our century, represents the march of civilization and the progress of science.

OLD BRIDGE, DARTMOOR.

Nevertheless, in its primitive form, the simple arch, the imitation of the "bow in the cloud," is most picturesque. Who of us, when in the country, can resist the mute appeal of the old time-worn bridge of slabs or crumbling masonry, ivy-clad, hoary in its honoured age, a monument of the centuries long ago, cemented by the lichen and the moss, above the stepping-stone on which paused our ancestors, who tramped "o'er moss and fell?"

It would be interesting to trace the evolution of the bridge; and what a history might be written of bridges! We could recount the exploits of the brave Roman, the defence of Lodi, and the desperate struggle on the, lately fallen, Devil's Bridge in Switzerland. And what poetic fancies, too, are conjured up by bridges ! A trysting spot, a place for "kissing toll," and one on which the ghastly heads of traitors were displayed to wither into grinning skulls. A lounging spot for gossips, the beloved of the antiquarian and the artist. Every phase of human nature may be found associated with the bridge ! There are sermons in the moss-grown stones, and what tales they could tell us ! Whose hands raised the old three-arched Gothic structure at Croyland, in the palmy days of the monks, in 860 ? Who erected the old slab-bridges of Dartmoor ? The Romans built timber-bridges in England, and their viaducts and aqueducts on the Continent remain to this day, but Dartmoor can show older viaducts.

A curious society, called "The Brethren of the Bridge," was established in France, and we know the Romans made the priests their bridge-builders, hence the title *Pontificus.* The Moors also excelled in bridge architecture, as Cordova bears witness. The Brethren of the Bridge in later years—in the twelfth century—took upon themselves this phase of engineering. The society protected the traveller who was liable to be waylaid in crossing the river, when his danger became imminent. The Brethren built and assisted in building bridges and houses of refuge ; and on the Durance they began their excellent work. The bridge at Avignon was constructed by a shepherd named Benezet, who on one occasion had a vision, in which he was commanded to go and build a bridge over the Rhone.

He complied, but could, unfortunately, obtain no assistance until he applied to the kindly Brethren, who fell in with his views. The bridge was erected and finished in 1188. But like other great engineers, the faithful Benezet did not live to witness the completion of his splendid work. He died in 1187, and was canonized. The bridge had eighteen arches, and was nearly forty-six feet high.

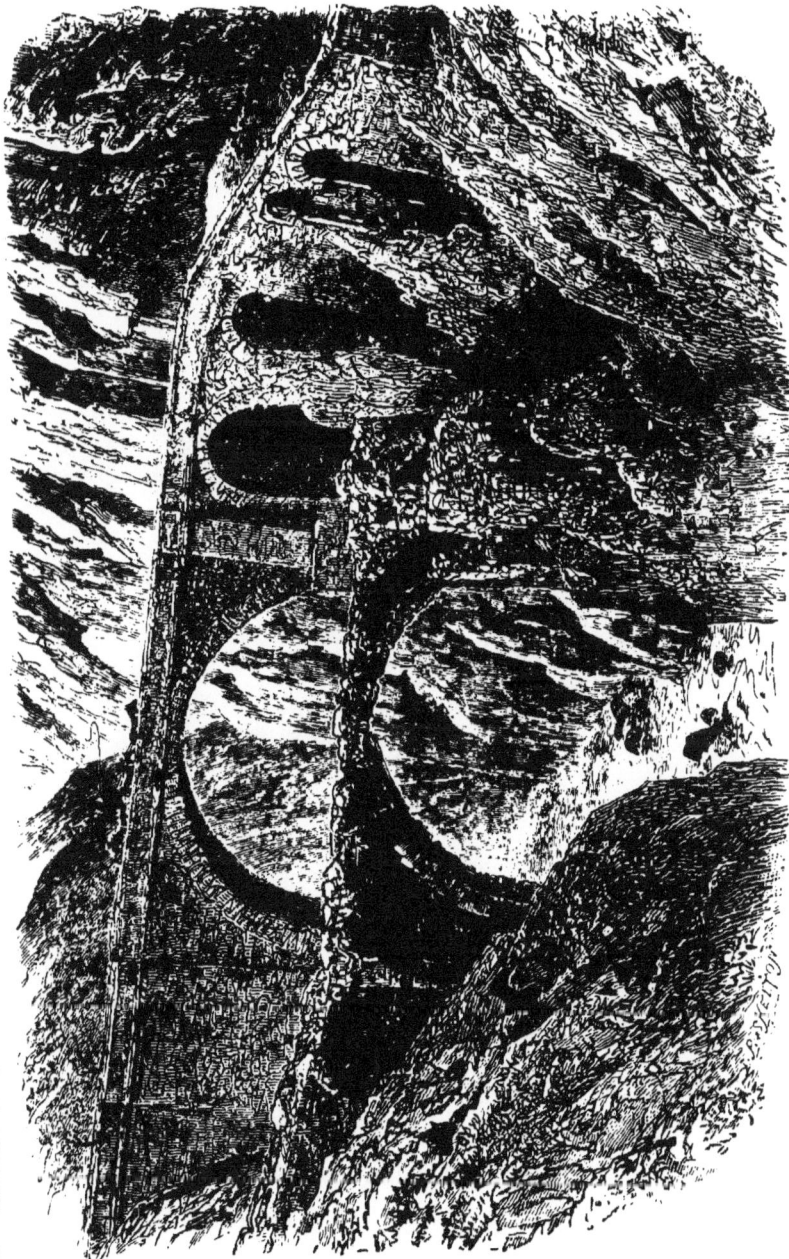

THE DEVIL'S BRIDGE, ST. GOTHARD PASS.

Croyland Bridge was the first stone bridge in England in 856-860. It was rebuilt in the fourteenth century, about 1380. The three arches unite in one—emblematical of the Trinity— and it is a curious development, inasmuch as only foot passengers can use it. The approaches are exceedingly steep, with steps, so no vehicles can cross. The streams it used to

REMAINS OF BRIDGE AT AVIGNON.

span are dry, and its foundations are in Lincolnshire, Cam- bridgeshire, and Northamptonshire. It is a curious relic.

Not until after old Croyland did Thames boast a bridge. In 993 a wooden structure over the river near St. Botolph's Wharf was erected. The river had been previously crossed by a ferry, worked by one " John Overy," who made it pay very well. The name of his daughter, Mary, is still remembered. She was a very religious as well as a very beautiful girl; and her charms added to her father's gains, made her an object of considerable interest.

She was beloved of a youth who anticipated the savings of the ferryman with much anxiety. Overy seems to have been an eccentric individual, for he on one occasion feigned death, anxious, as was Grimaldi the elder in this century, to ascertain the feelings of his domestics when his loss should be made known. He pretended to. die, and had himself laid out, in the expectation that his servitors would fast religiously. Only Mary was aware of the plot; but to her alarm, and to the inexpressible disgust of the "corpse," the unfeeling domestics broke open the larder, or buttery, and helped themselves liberally!

This was more than the "deceased" could bear. He flung off his winding-sheet, and hurried down in his cerements to confront them. Unfortunately, his appearance did not sufficiently appeal to his people. They regarded him as the Prince of Darkness, and one of them boldly attacked him. A blow despatched the unfortunate ferryman, who was then dead in very truth—an example of the old proverb, "Mocking is catching."

The news quickly spread. Mary was an heiress, and the gallant youth came to her, riding with headlong haste to win her and her fortune. The more haste, the worse speed. The "gallant" threw his horse, fell, and was killed; and poor Mary, not having any use for her wealth, dedicated it to the Church, and retired to a convent. The church she founded still bears her name, St. Mary Overy; but, as Mr. Smiles pertinently remarks, this may be only another form of "St. Mary o' the Ferry," and the ferryman a simple "John o' Ferry." A poor Jack, we may say; and the story is, perhaps, all tradition.

However, the wooden bridge was destroyed in 994, when the Danes came up the Thames; but it was subsequently repaired, and proved an obstacle to our old acquaintance, Knut, or Canute, who took the trouble to cut a canal from Rotherhithe to Chelsea. So the wooden bridge lived, but was again and again destroyed by fire and flood. At length Peter, of Colechurch, made up his mind to build London Bridge of stone, and set about it manfully.

A very old ballad, which used to be sung to the writer, when a child, by his nurse, commemorates this destruction of the bridge. The ballad was chanted to the tune of " Rob Roy Macgregor, oh," as I afterwards recognised it. It began—

> " 'London Bridge is broken down,
> Gran',' said the little dear ;
> ' London Bridge is broken down—
> Gay laydie ! ' "

The verses then proceed to detail all the material for rebuilding it, "lime and stone" being the last suggested. "Wood and straw" had been rejected with other such flimsy adjuncts; and London Bridge was presumably rebuilt, on "packs of wool," which, even to my childish mind, seemed a very unstable foundation.[1]

CHAPEL OF ST. THOMAS À BECKETT, LONDON BRIDGE, 1205.

The "old" London Bridge was begun in 1176, and was of stone, defended by "starlings," or coffer dams, which, originally built to defend the piers, were never removed. There were twenty arches. But Peter of Colechurch died in 1205, and although a Frenchman applied to finish the work, three merchants took it in hand, and finished it themselves. In

[1] This has since been explained to mean the proceeds of a tax on wool, which provided the funds for the building of the bridge.

after years houses were built on the bridge. A Gothic chapel was also erected, and dedicated to St. Thomas, in which was laid the body of Peter the bridge builder within a pier of the structure he had designed. Unfortunately this chapel was pulled down in 1760. It had been, of course, suppressed as a place of worship, but was in existence as a shop, from which a stair led to a curious fish-pond, formed in the coffer-dam underneath.

There was a tower at each end of the Bridge, and near the centre of the drawbridge. Nonesuch House, a quaint and curious structure, was erected on it. The houses were destroyed in 1666 by the Fire, but rebuilt, and were finally cleared off in 1755. But old London Bridge remained till 1832 a monument of Peter's skill.

Numerous stone bridges were built in England after Peter's time, but their construction need not be dwelt upon. Meantime the roads remained in a very bad condition, the means of locomotion being — save on horseback — very rough also. The unfortunate occupants of the carts and carriages were so terribly shaken about, that the rivers became " highways " instead of the roads.

Of all these " silent highways," the Thames was the most frequented. History tells us of the pageants and processions which passed along its surface, of the numerous ferries still in name surviving, as the Horseferry Road at Westminster witnesses. Hunger-ford is another, and there were others lower down. The same practice obtained in other parts of the kingdom, for bridges were erected but slowly.

The narrow dimensions of the roads, as a matter of course, restricted the bulk of the articles carried over them. The narrow ways, therefore, limited the supply, save by increased numbers of packs; and so, many teams of pack animals made their way into London, while other consignments came by water. The necessity for this communication opened up the rivers, and caused the cutting of numerous canals, by which the country districts could be tapped. As will be seen, road

LONDON BRIDGE IN THE SIXTEENTH CENTURY.

l-gislation improved matters in 1285, and, by degrees, the engineer, as we know him, arose.

But we do not intend to join hands with the modern engineer just yet. We shall rather put him aside, thrust him back. We wish to pry about us, and find out the old, old paths, very crooked and "hard to find out all alone," as the ballad says. We would rather revert to the time when the monks of Walsingham took the Milky Way as their guide to the shrine there, and palmers' feet pressed the stony road by Fakenham and Brandon.

Out upon the engineer who has spoiled so much of our romance in these days ! We will not hurry to meet him on the road. We will rather linger while the phantom coaches, and the ghosts of highwaymen, pass silently along the king's highways, rough and muddy as they were, even less than a hundred years ago.

LONDON BRIDGE IN THE NINETEENTH CENTURY.

HIGHGATE ARCHWAY GATE, 1825.

CHAPTER II.

ROADS AND TRAFFIC IN EIGHTEENTH CENTURY.—HOUNSLOW
HEATH AND HIGHWAYMEN. — "ON THE ROAD."—THE
PIONEERS OF ENGINEERING—EDWARDS, AND METCALF:
"BLIND JACK."

THE first Act concerning roads was passed in 1285,
as we have seen, and the first tolls were levied in
1346, on carts or carriages travelling from St.
Giles-in-the-Fields to Temple Bar, which route,
we read, was much choked by bushes. The ruts and holes in
the streets had, in those days, to be filled in with faggots on
state occasions.

Henry the Eighth compelled the parishes through which
the roads passed to maintain them, and Charles the Second
established the toll to keep the roads in repair. These tolls

were never relished. In the olden time the sovereign used to maintain the roads which belonged to the people, who now rebelled at the impost. The notorious Rebecca Riots ensued, during which the turnpikes were destroyed in South Wales.

In the days when nearly every one rode on horseback we find the commercial traveller with his packages in bags hanging from the saddle—hence the term "bag-man," which clung to the "knights of the road" for many years. Now these knights are commercial gentlemen, in carriages ; but in those early eighteenth century days they rode, as coaches were very slow and tedious. Vehicles made but little way in public estimation, because of the narrowness and bad repair of city streets. The "big people" by degrees migrated to the western end of town, where was more room for "coaches." From the beginning of the seventeenth century the use of the "coach" became more frequent, and about the middle of the century the stage-coach, or rather we should say stage-wagon, came in.

But, of course, owing to imperfect roads, the dangers of travelling were numerous. The rivers washed over the roads, and at times fatal results followed. On many occasions all traffic was suspended by rain, and travellers had to remain on the road, or return till the floods subsided.

The difficulties in getting about even as late as the end of the eighteenth century, would seem incredible to us at present if we had not so much contemporary testimony to turn to. Travellers in the earlier part of that century were still compelled to ride on horseback, generally, as carriages were so frequently overturned, or stuck fast in the ruts or in the mud. Even Royalty suffered. When Prince George of Denmark set out from Windsor to reach Petworth, he left the castle at 6 a.m., and did not accomplish the forty miles in less than fourteen hours, during which time the occupants did not quit the coaches, "save when we were overturned, or stuck fast in the mire ! "

In 1743 bitter complaints were made of the condition of

the roads even near London, notwithstanding the tolls levied to make good the highways, which in summer were deep in dust, and in winter very Sloughs of Despond, deep enough almost to engulf Christian himself. The road to Kensington was simply impassable in consequence of the mud, and though the Turnpike Act was passed in 1755, not much improvement ensued for some time. Save one or two roads, Mr. Young writes in 1767 in his six weeks' tour :

" It is a prostitution of language to call them turnpikes," and he inveighs in strong terms against the infamous conditions of the roads called "turnpikes," but they did not extend beyond Grantham, up to which town on the north side, the old causeway still existed in 1739. The reader will then perceive the immense difficulties of transport, and yet the tolls were resisted. In Yorkshire the crier actually called the people together publicly to bring hatchets and axes to destroy the turnpikes at midnight. These directions were carried out, and, notwithstanding the resistance of the military, many toll-bars were destroyed. Between the years 1760 and 1764, some four hundred acts of parliament were passed to amend old or to make new roads.

To tell of Hounslow and its highwaymen scarcely lies within our province in this work, but, as a romance of the highway, a passing reference should be made to some of those so-called "heroes of the road," who were, in sooth, but miserable ruffians. That there was some politeness inherent in the French valet Du Val, is not unlikely ; and his celebrated dance upon the heath with the fair occupant of the " coach," would be romantic if he had not exacted a ransom from the lady's father, or husband. Dick Turpin, whom Ainsworth has elevated into a hero, never rode to York as described ; he was a mean robber, who with his gang haunted Epping or other convenient spots, and plundered the weakest. Whitney was another "hero" also a butcher, as was Turpin. This worthy had some traits of generosity, for on one occasion, having robbed an old gentleman, he (the traveller) begged for a small sum as he had far

c

to go and no means to obtain money on the road. Whitney opened the bag and bade his victim help himself. Mr. Long —that was his name—took a handful of the coin, whereupon Whitney remarked, "I thought you had more conscience, sir," and rode off with the remainder.

No doubt these gentry were somewhat picturesque in attire : "the gold laced cocked hat and scarlet roquelaure"; the silver pistols in the deep pockets ; the "speedy chesnut mare" of "the reckless rascal" skimming over the heath, or leaping gates and brooks by night to meet the heavy coach—the canary-coloured "Comet"—with its fourteen passengers and natty coachman ; the guard, as the name implies, armed with blunderbuss and pistols. The highwayman rides up, the whole party is transfixed, the "silver pops" are pushed through the glass, and in a few minutes the passengers are bemoaning their property, while the blunderbuss at last is discharged in the direction of the moon.

Stories by the dozen could be related of these escapades, not always performed without a contest or a capture. Sometimes a "hero" was shot or captured by persons in authority, who travelled on purpose. Sometimes the solitary equestrian escaped by simulating another of the fraternity of thieves, and by passing himself off to his arrester as a partner in crime !

Jack Hawkins used to steal the mails, and even stop coaches in Lincoln's Inn fields and Chancery Lane. After dinner at Brentford, he passed on to the Heath and "emptied the mail." This man and his associates were well-known ; and bribed ostlers, maids, toll-men, and even constables to give information, or to assist them to escape. Rann, known as "Sixteen String Jack" because of the ties at each knee, was taken and executed for robbing on the Heath ; and Hawkins was hung in chains near Hounslow for his misdeeds ; "Jerry" Abershaw added to the real terrors of the road. . . .

But the inveterate opposition continued. George the Second had a law passed, making it felony to destroy a toll-bar ; and on the other hand, the people sent up a petition against the extension of turnpike roads, for fear the country traders would outsell and undersell the London dealers and market gardeners ! Again, stage-coaches were objected to on the ground that the trades of saddlers and spur-makers would be ruined if progression on horseback were relinquished. Road-side inns would be injured, and, as a "clincher," the Thames traffic would certainly suffer !

In the middle of the last century the mails were carried on horseback, or in unprotected carts, more slowly than any other consignments. The consequence was that robberies were frequent—even if the post-boys themselves were not in the secret. Highwaymen increased, and even after 1784, when Mr. Palmer's new stage coaches commenced to run, the attacks upon the mails were frequent. Many romantic incidents were connected with travelling in those early days of "Flying Coaches" and mails. The engineer had not yet made his presence felt, although the common sense of the community was beginning to display itself. To travel by coach in those days was an undertaking for which people prepared themselves by making their wills and loading their pistols. A few anecdotes will illustrate the complexion of the times, and the inconvenience to which passengers were subjected in the company of the ailing and the insane ; squalling infants, and untidy women ; bullying men and coarse ruffians who had paid, and intended to go. The atmosphere of such stages must have been absolutely poisonous, and when, as sometimes happened, an extra passenger was introduced, no one would quit the coach for fear of being left behind.

What are we to think of our metropolitan road-menders of the time when even the Oxford road possessed, in winter, or in bad weather, only one passable track, less than six feet wide, and "eight inches deep in fluid sludge !" At the end of the eighteenth century a coach stuck in the mud, as a matter of course, or if it reached Hounslow Heath it was again stopped by some of the gallant highwaymen—Du Val or another—and the passengers were plundered.

Nor were the attentions of these gentlemen limited to Hounslow Heath or Shooters' Hill. The Knightsbridge road, which was carried across the West-bourne stream, was a favourite haunt of highwaymen. There are records of the "Exeter Fly" having been stopped by a masked horseman, who demanded the travellers' money, and added that haste in delivery would be appreciated.   ·

Notwithstanding the terrified compliance of some, one individual closed with the robber, whose pistol did no damage, and a wrestling match in the mud ensued. The coachman attempted to drive away but again stuck, and when the guard arrived with assistance, he found the passengers binding the highwayman, who was presently hanged. Readers of Smollett will remember many episodes of life on the road, and any one desirous to read others can refer to Mr. Timb's "Romance of London," "The Coaching Age," and such interesting annals of our roads of the past. Travelling in daylight was then bad enough, but in a fog it was terrible work. Then the coaches formed in line, and followed each other through Hounslow, whither "eight mails" went before they separated to Bristol, Exeter, Staines, Bath, Gloucester, etc. Mr. Wood recounts a mistake in taking the road which was rectified by the accident of one coachman recognising the voice of another, saying, "Charley, what are you doing on my road?" He was wrong, though, and "Charley" was right. The speaker had missed the road in which gallows, and bodies hanging in chains, were features.

In those "good old times," as we often say, people travelled to Paddington up the wooded Edgware Road by the Tybourne in company, at intervals, for fear of attack! Even Piccadilly and the Parks were not free from footpads, who demanded money in broad daylight.

But the day was approaching when the turnpike road would become an easy, pleasant route, free from accumulations of mud, and deep holes into which a barrow could disappear; void of boulders which broke the axles or lamed the horses;

and maintained by well-paid labour. The engineer was approaching! Up to Smeaton's time no one believed in the profession of "engineer," and the man who would turn his hand to road-making, as an occupation, was regarded much as a crossing-sweeper is now.

Therefore no one of any talent took up the business. The mending and making of highways was considered derogatory.

JOHN SMEATON.

Bridges were regarded differently. Many eminent men, as we have seen, devoted themselves from charitable motives to building the bridges, but no one attempted to get to heaven by the road! Therefore, as in the case of steam discovery, the work fell into the hands of poor men: men of practical knowledge, uneducated. Two instances of this will suffice to close this portion of our subject before we come to the wonders of our own century.

These apostles of the Bridge and of the Highway were, respectively, William Edwards and John Metcalf, and their history is a romance.

William Edwards, born in 1719, was a strong willed lad, son of a farmer, and imbibed a taste for engineering by putting up the walls or stone fences in his district. This taste grew upon him, and he took every opportunity to indulge it. From an amusement, or at least a leisurely occupation, this building craze became a business. Edwards was employed by acquaintances to mend their fences, and so by practice and study he became a very expert mason by the time he had reached his twenty-first year.

His first real advance was in the building of a mill, which was a success, and consequent on this feat he was required to build a bridge over the Taff. He had never received any instruction in bridge-building, and he undertook the task with some misgivings. He proceeded, however, and succeeded in raising a bridge of three arches of stone, but the river Taff, which had never been so spanned, rebelled; it rose one stormy night, and after filling the arches of the new bridge with driftwood, and other floating substances, pressed against the piers, and, after a while, like Samson, succeeded in pulling down the structure on itself. Then it ran impetuously on, and next day the stream, somewhat subdued, flowed over the ruins of Edwards's first bridge.

But if the Taff was resolute, so was Taffy. Though he could not help feeling the blow, he did not despair. By the terms of his contract he was compelled to keep the route open; so he again set to work, and this time determined not to give the river any opportunity to overthrow his piers. Edwards designed his bridge of one arch only, of one hundred and forty feet span—the celebrated Rialto is only ninety-eight feet.

Such an undertaking by a self-educated "engineer," gave rise to comment. Would he succeed? He determined that he would. He built the arch and was engaged in the erection of the abutments, when their upward and lateral pressure forced out the keystone and centre of the arch, and the bridge once again tumbled into the glancing river.

This second disaster nearly ruined him, but his friends rallied round him, and he made a third attempt. In this he succeeded. He took off the pressure by drilling, or leaving, in each haunch of the bridge three holes of varying diameters, the largest (six feet) inside, or lowest. This plan succeeded. Pont-y-Prydd, the "Bridge of Beauty," was finished—a beautiful rainbow structure, but unfitted for the passage of vehicles. It is so steep that a drag was employed to check the descent

BULL AND MOUTH INN, ALDERSGATE STREET, ABOUT 1828.

of a carriage, a lad riding on the counterpoise for his trouble in attaching it.

Pretty as it is, Pont-y-prydd is not so useful as Edwards' subsequent successes. A new bridge has superseded it now for traffic, but the Bridge of Beauty remains a monument of a man of resource and talent, who subsequently added greatly to his fame. The celebrated bridge was finished in 1789, and still stands.

Of Edwards's subsequent career in bridge-engineering there is ample testimony. He constructed several handsome bridges in Wales, varying his work by preaching—for he was a rigid Calvinist in his tenets —and building workshops. His countrymen, who had flocked to see his first bridge, had every reason to be grateful to him, not only for his public works, but in consequence of his private benevolence. Edwards was a man who made his own way; he was the architect of his own fortunes, and as the revivalist of the art of bridge-building in this country, he is fairly entitled to a niche in the Temple of Fame.

Edwards left four sons and two daughters : three of the former were brought up in their father's trade ; the fourth was shot in the American War. David Edwards turned out a skilled bridge-builder, many of his constructions remain in South Wales. "Bridge-building and farming," says Mr. Malkin, "seem to be the hereditary employments of the family." [1]

The career of John Metcalf, known as "Blind Jack of Knaresborough," is a curious one. An old biography, published in 1795, supplies most of the information regarding him, and he certainly seems to have been a wonderful man in his way. It appears that Metcalf was born in August, 1717, and was sent to school at four years old. He continued to study until a virulent attack of small-pox deprived him of sight, but not of his energy. He speedily found his way about the town, and in the course of a few years was able to associate with other lads in their games and pastimes, even to the extent of birdsnesting and riding. He also learned to swim, and was instrumental in saving life by his dexterity.

Notwithstanding his blindness, young Metcalf managed to get into a few scrapes ; he and his companions played many tricks, but they also acted as benevolent imps. On one occasion Metcalf was the means of recovering a quantity of yarn which had been submerged by sudden flood. He managed to hook the yarn beneath the surface of the deep stream, and, greatly to the surprise and to the greater satisfaction of the owner, the lad and his assistants recovered the greater portion of the yarn.

He did not confine himself to these acts, however. His practical jokes and retaliation upon anyone who had injured him became celebrated. He would pen a whole family in the house, exclude the daylight, and deluge the rooms with water; liberate a neighbour's sheep, rob a rookery, or descend a chimney, naked, to open the house-door for a friend ! On this last occasion he was compelled to leave his clothes on the roof, as the flue was too narrow to admit him dressed; and rain coming on Metcalf was compelled to remain unclothed till his garments had dried. This feat we are told was considered a very kind act towards a friend.

His taste for music, and his ability as a violinist obtained for him the appointment of fiddler at Harrogate, and subsequently he led a small band there. Metcalf was fond of sports—such as racing, cock-fighting, and hunting, in which he freely associated with the gentry, with whom it seems he got on very well, and was addressed as the " Blind Huntsman."

An anecdote of his marvellous knowledge of locality, in spite of his infirmity, may be related here. About this period Metcalf, at an inn in York, encountered a gentleman who required a guide to Harrogate. The landlord, carefully concealing the fact that Metcalf was blind, recommended him, and informed the young man. Metcalf agreed, and though no turnpike road then existed between York and Harrogate, the Blind Huntsman led the way unerringly through the Bar. through Poppleton Field to Hessay Moor, and so over Skip Bridge. The blind guide replied to the stranger's enquiries regarding the features of the country with readiness, and the gentleman had no suspicion of him.

A sudden turn in the road at the end of the moor in no wise disconcerted Metcalf, who held on his way. Further on their road was crossed by another track, and proceeded alongside a brick wall of Allerton Park. There was a road leading out of the Park, opposite to the gate on the Knaresborough road, and Metcalf was afraid of missing it ; but a puff of the east wind through the gate gave him his cue, and he turned his steed to the opposite gate, which he tried to open *at the hinges*, but quickly laid the blame on his horse.

The gentleman beside him soon opened the gate, and Metcalf concealing his face, rode on in the gathering darkness, which, of course, was all the same to him. His companion, never suspecting Jack's infirmity, would have stopped to drink with him, but Metcalf declined, and went on. By marvellous skill, and cool self-possession, the guide made his way over the causeways and narrow paths, and by skilfully veiled questions managed to ascertain from his companion their distance from certain twinkling lights which the gentleman had remarked upon.

At length Harrogate was reached in safety, and at the hotel Metcalf's companion remarked to the host that the guide looked as if he had been drinking freely, " his eyes seemed so queer ! "

" Eyes ! " exclaimed the landlord. " Do you not know that he is blind ? "

" What do you mean ? " cried the traveller.

" I mean that he cannot see ! "

" This is too much ! " exclaimed the gentleman ; " call him in. My friend," he continued, when Metcalf entered, "are you really blind ? "

Metcalf assured him of the fact.

" Had I known that," replied the gentleman, " I would not have ventured with you for a hundred pounds ! "

" And I, sir, would not have lost my way for a thousand," retorted Metcalf."

This is not the place to detail any more of the adventures of this singular man, though the tale of his exploits, his elopement with his intended—the betrothed of another—would have suited Lochinvar. Metcalf served with the army of General Wade, and being a fine, tall and merry fellow, "Blind Jack" attracted attention. He had continued to play to the Yorkshire assemblies, and the manner in which he traversed the country, destitute of any proper roads, was marvellous. In London he played the fiddle, and so kept himself. He found a good friend in Colonel Liddell, of Ravensworth, whom he fairly astonished by walking the distance from London to Harrogate, two hundred miles, on an unknown road, quicker than the Colonel performed the distance in his coach. A more instructive comment upon the state of the roads at that time could not be adduced.

After his campaigning, which ended up at Culloden, Metcalf returned home, and began at length to settle down with his wife and family. In 1751 he "set up a stage-wagon"; in other words, he became a carrier, and made the journey between York and his native place twice a week in summer and once a week in winter. This occupation he continued until he undertook to make a portion of the new road between Harrogate and Boroughbridge.

It was "about the year 1765," says Mr. Smiles, that the Act of Parliament for this road was passed. A person of the name of Ostler was appointed surveyor, and Metcalf, who knew him, was interested in the proposed thoroughfare, no doubt because his long and wide experience told him how necessary good and properly kept roads had become. Metcalf, "falling in company" with the surveyor, agreed with him to construct three miles of the highway, the material to be drawn from one gravel pit.

The "engineer," as he may now be called, found men and materials. He drew the gravel from the pit, built a shelter for his men, carried out the daily meat himself, and by such energy and self denial, keeping steadily at work, he and his gang finished the space of road, for which they had agreed, to the entire satisfaction of their employers. This was the first attempt at road-making in late years, and we may regard Metcalf as the pioneer of the highway engineers.

Once launched upon this kind of work, Metcalf struck out

manfully. The road required a bridge, and a number of gentlemen met at the inn at Boroughbridge to consider the matter. Metcalf joined them; estimates varying considerably were put in, the road-maker tendered his services to the surveyor, and though Metcalf had never made a bridge and scarcely seen one, his name was sent in to the company.

Summoned before the gentlemen at the "Crown," Metcalf was questioned as to what he knew about bridge-making. He at once dictated his plan, gave the dimensions of his proposed bridge, indicated the whereabouts of the materials, etc., and so pleased the assembly that he was entrusted with the building of the bridge, greatly to the annoyance of the other tenderers. They attempted to thwart him by refusing to sell him the stone he required ; but the engineer was not without resources : he proceeded in another direction, found a quantity of stone, and soon had it conveyed to Boroughbridge, where the arch was promptly erected.

Pleased with his success, he again undertook a job of road-making, and in going over the ground he and his men discovered an ancient Roman causeway, the stones of which were utilised. But presently the workmen came to a piece of boggy land, over which the surveyor believed the road could not pass, although that was the proper direction. He was for turning aside, but Metcalf declared that he could make the road over the swamp, the nearest way.

"Well, then, if you can, you shall be paid as if the road had gone round," said the surveyor. "Jack set about it," says the chronicle; "he cast the road up, and covered it with whin and ling, and made it as good, or better, than any part he had undertaken." For this job Metcalf received four hundred pounds, with eighty of which he purchased a house and some land, which tenement and rights of commerce he soon afterwards sold for two hundred pounds.

Thus fortune accompanied him. His "luck," as people termed it, was simply the result of his thoughtfulness and shrewdness. His busy brain planned out all these schemes, and being an excellent judge of character by hearing and perception, Metcalf was enabled to protect himself.

Another road was now required, six miles in length, and for this Metcalf tendered. He obtained the contract, and with his accustomed energy set to work. He completed his task, and received twelve hundred pounds for it. Then work flowed in, and the road-maker tramped many a mile to perfect his designs or to verify calculations.

These calculations were all made in his head, and very rapidly as well as correctly. His greatest triumph was on the Wakefield and Austerland road, the portion of it for which Metcalf tendered having been carried over deep marshes. He complained, but was requested to proceed, to dig out the marsh and make the road on the solid beneath.

But this Metcalf declined to do; the labour he ascertained would be immense, and if such a hollow way was made, it would certainly be blocked in winter by drifting snow. The engineer tried to persuade the trustees, he could not, so he took his leave, saying that he intended to try his own plan first; if that failed he would try theirs at his own expense. This arrangement was agreed to. He commenced at six places at once; he levelled the path, bound heather in bundles and laid them on the route in squares of four, placing another upon each square, pressing them well down. Then he provided broad-wheeled carts, and laid on stone and gravel. The carts went on amid fears and derision, but returned amid cheers, which were deserved, as this piece of the road required no repairs for twelve years. In this method he anticipated George Stephenson.

In 1778 Metcalf lost his faithful wife, the daughter of the innkeeper at the Royal Oak at Harrogate, at the age of sixty. He continued his profession until he was seventy, and joined his married daughter in the cotton-spinning trade. When he gave up work, in the seventy-sixth year of his age, he amused himself by dictating his exploits, which have been published, and from which book, with others, this notice has been compiled.

He enjoyed the perfect possession of his mental faculties,
"and enjoyed the company of his friends until the month of
April, 1810, when, on the 27th of the month, he died, in the
ninety-third year of his age—eighty-seven of which were spent
in "perpetual darkness. . . . He left four children, twenty
grandchildren, and ninety great, and great great, grandchildren."

By this man of undoubted talent the road-system of England
was immensely benefited.   The short account of his life which
we have given will convince all readers of his energy and
determination, and that John Metcalf, in his blindness, was
largely composed of the stuff of which heroes are made.

TEMPLE BAR, IN 1830.

OLD HOUSES, EDINBURGH.

CHAPTER III.

OLD SCOTCH ROADS.—GENERAL WADE.—THOMAS TELFORD AND HIS WORKS.—IRON BRIDGES AND NEW WAYS IN SCOTLAND.

T was during the rebellion in Scotland in 1715 that the British Government awoke to the importance of making roads, and ten years later General Wade, whose name is still associated with a rhyme of somewhat Hibernian tendency,[1] was appointed to survey and report upon the roads in the Highlands.

He sent in his report, and in consequence he was instructed to proceed with several regiments to Scotland, and make the roads, so that in any future eventuality troops and war material might be easily transported through the country. These precautions were justified in the '45; and forty years later their course extended over nearly eight hundred miles.

Wade appears to have been a most energetic engineer. Pennant calls him "another Hannibal," who found his way

---

[1] "Oh! had you only seen these roads before they were made,
You'd lift up your hands and bless General Wade."

through rocks supposed to have been unconquerable. Whether these roads can be termed good, or "highways," in any other sense than that of elevation, is a matter of opinion; but they were boldly constructed, and served their purpose.

While Wade's troops were digging, blasting and levelling the Highland tracks, an infant was born in Eskdale, on the banks of the Esk, at Westerkirk, who was destined to revolutionize the art of road-making, and to hand down his name to posterity as a benefactor to trade and commerce. This child was the son of a poor cotter and shepherd in the pastoral district which lies between Dumfries and Roxburgh, whose name was John Telford. He died when his son, destined to be so famous, was only four months old.

We have it on the testimony of Thomas Telford himself in after years, that he "recollects with pride and pleasure" his native parish of Westerkirk, where he "was born in the year 1757," and there he went to school in the winter, when out-door occupations did not so greatly demand his services. During the other portion of the year he was employed by an uncle, and thus he managed to assist his mother. In the account given of him by Mr. Smiles it is stated that Telford was a merry, laughing lad, that the farmers in the neighbour-hood were very kind to him and his mother after the death of his father.

As Tom, or "Tam," grew up, he was sent to learn the trade of a mason, and we read that he was sent away to Loch-maben. He did not remain long in service, however; the ill-treatment to which he was subjected determined him to run away, and one day, after some few months' absence, master "Tam" turned up at the little cottage in Eskdale again.

His sudden appearance caused his mother some anxiety; and after she had embraced him, she likely wanted to know why he had paid her such an unceremonious visit. Her con-sternation may be imagined when she learned from her son that he never intended to return to his master any more; but would just remain in the valley for the present.

It may be supposed that his old school-fellows, the Malcolms, of whom in his autobiography he speaks so highly, had quitted the neighbourhood, or perhaps they might have assisted him. "The eminent brotherhood of the Malcolm family" were apparently not applied to, and the lad was again sent to a mason—this time by his cousin—to perfect his experience in the trade.

While thus engaged, Telford did not neglect his education. He read, and practised writing, gained considerable experience, and then at twenty-five, he says, he considered himself to be master of his art—"as practised in the county of Dumfries." This was not apparently very high-class work, for the houses

WATERLOO BRIDGE, LONDON.

were small, but it is recorded of him that one of his first acts, when he could hew and chisel well, was to cut an epitaph to his father's memory, thus :—

"In memory of John Telford, who, after living 33 years an unblamable shepherd, died at Glendinning, November 1757."[1]

There is something extremely touching in the affectionate simplicity of this epitaph, "an unblamable shepherd" is most artistic and filial. There is a touch of poetry about it also, and it is known that Telford wrote some poems, notably one on "Eskdale," which is printed in his autobiography.

At twenty-three years old he started for Edinburgh, and in that ancient British city, he says, "commenced the splendid

[1] Smiles' "Lives of the Engineers."

D

improvements which have since extended in every direction."
Two years later he rode to London on a horse belonging to
Sir I. Johnstone, lent him for the purpose of being taken
thither to a relative of the baronet.   An anecdote is told by
Mr. Smiles that his cousin Jackson lent Tom his buckskin
breeches to ride in, "but 'Tam' always forgot to send me back
my breeks," added the lender.

In London, Telford was employed on Waterloo Bridge and
Somerset House, and subsequently in Portsmouth, and the
works, "under my superintendence," having been completed
in 1787, the "architect" went to Shrewsbury Castle, at the

SHREWSBURY CASTLE.

invitation of Sir W. Pulteney, formerly a Johnstone, of Telford's
native parish, who changed his name on his marriage to Miss
Pulteney.

At Shrewsbury Castle Telford remained, and subsequently
was made county surveyor.   It was while at the Castle that the
discovery of the old Roman city was made, and Telford was
employed to carry on the excavations on the authority of
Mr. Pulteney.   The Roman "baths" were thus opened up,
and Uriconium was revealed.   The chambers are really
"hypocaust apartments in a tolerably good sized dwelling-
house."   Another interesting employment was stopped by the
negligence of the local vestry.   The church-tower was in

danger, and Telford would have had it taken down but for the obstinacy of the authorities.

So it happened that when the bells chimed the tower fell, and buried a considerable portion of the edifice. That the danger was imminent may be gathered from Telford's remark when they wished to discuss his motives : " If you wish to discuss anything besides the alarming state of the church," said he, " you had better come to some other place, where there is no danger of it falling on your heads." The tower collapsed a few days after this warning.

Telford had been appointed to the post of county surveyor, and engineer to the Shrewsbury Canal, in 1795, and in those capacities he seems to have bethought himself of the use of iron for bridges and aqueducts. In the use of this metal for aqueducts, Telford was probably first, but, as regards bridges, he was not a pioneer of the use of iron. His attention was drawn to its use because, as Mr. Smiles says, " Shrewsbury is situated in the immediate neighbourhood of the Black Country, of which coal and iron are the principal products."

Nevertheless, our energetic engineer was not the introducer of iron bridges. We have to turn back twenty years to find the first iron bridge, although attempts had been made in the middle of the eighteenth century to build a bridge of cast iron. It was never completed ; indeed it had been scarcely commenced when the attempt was abandoned. To the French this new departure in bridge-building is due, although their engineers found it necessary to relinquish the attempt for several reasons, chief amongst which was the price of the metal. To this economical view may be added the difficulty which then existed in producing sufficiently large girders of cast iron ; for at that time the iron-workers could not turn out the castings as they did a few years later, when the manipulation of iron was better understood.

Naturally, the first practical proposal for an iron bridge came from an iron-worker, Mr. Darby, of Coalbrookdale. There, as elsewhere, the increasing traffic, consequent upon the improvement of the roads, was a source of embarrassment at the ferries and fords. Disputes arose, delays occurred, and as the ferryman was a dictator, yet perhaps open to monetary persuasion, we can imagine that some passengers were compelled to lose both time and temper in the transit. The improvement in the highways, which had already demanded stone viaducts, now brought pressure to bear upon the iron district, and iron was considered the most fitting as well as the " handiest " material to deal with.

Mr. Darby therefore, in 1776, desired an architect of Shrewsbury to prepare a design for an iron bridge. Darby himself superintended the castings in his " works," and the result was the erection of a rather elegant bridge, of a single arch, one hundred feet span. This, the first success in iron bridge-building, is therefore attributable, not to an engineer, but to the owner of the local iron-works, who was justified in the employment of the metal as fully as were the classic cobblers who declared that " there is nothing like leather."

But notwithstanding the undoubted advantages gained by the use of iron, it was not rapidly adopted as a material for bridges. No doubt the same influences, which had militated against the French engineers, were at work in England. Facilities were wanting, and it was reserved for a reckless political adventurer to bring forward the question once more and solve it, although he did not himself succeed in erecting a bridge.

To no less notorious an individual than Thomas, generally known as Tom, Paine, was the resuscitation due. In early life this personage had assisted his father, a member of the Society of Friends, in "staymaking," but Tom had a soul above corsets, and quitted the shop for a life of adventure at sea. After serving in a privateer he became a grocer and exciseman at Lewes, but was dismissed for keeping a tobacconist's shop—for at that time Civil servants were not permitted to " run " stores. He sent in a petition on behalf of excisemen, which brought him into notice, and an Inland Revenue commissioner gave him an introduction to Dr. Franklin, who advised him to go to America.

Paine reached Philadelphia in 1774, and became editor of the *Philadelphia Magazine*, and took prominent part in the many revolutionary discussions of the period, going as far as to advocate in a pamphlet the severence of the colonies from Great Britain. He also studied science, and when the difficulties of building a bridge over the Schuylkill river were discussed, this erratic genius sent in a plan for an iron bridge of a single span. He submitted his suggestions to the learned societies in Europe, and himself followed up his specifications to Paris and London. The bridge was cast, and was actually set up in Paddington, where it was much admired.

So far he had advanced ; scientific men and the public pronounced equally in his favour, and Paine, assisted by American friends, might have succeeded in a more useful sphere, had not the French Revolution carried him once again into the maëlstrom of politics, in which he was engulfed. He escaped from a French prison to America in 1802, where he remained till his death, most unhappy in his social relations in consequence of his irreligious views and continual intemperance. He died and was buried on his own farm.

He had quitted England in 1792, leaving a memory, his bridge, and debts behind him.

To satisfy the last, his creditors seized the iron—then valuable—of the new bridge on the Paddington Green, and then arose the idea of utilising it. There were the component parts, why not put them together in some place where they would answer their intended purpose ?

The manufacturers, with their iron on their hands, had some difficulty in looking about them and of moving in the matter ; but fortunately a gentleman of Sunderland, named Burdon, determined to bridge the Wear there, and made a design for an iron bridge to span the stream from one high bank to the other. The materials of Paine's bridge were already made to hand, and although the new Wear Bridge was not erected on

Paine's plan, the same iron was used, and the bridge was completed in 1796.

This is the curious history of the bridge which commanded
attention by the boldness of its design and for its span, which
at that time was the largest in England.   The dimensions are
given as follows : The arch commences ninety-five feet above
the river and extends two hundred and thirty-six feet across it,
to the opposite pier.   The light structure rises thirty-four feet
in the centre, thus the space underneath for vessels is at the
highest point one hundred and twenty-nine feet from the
water.

Mr. Burdon received much gratitude and gained some
celebrity for thus executing at his own expense such a benefi-
ficial work ; and then iron bridges began to come into vogue.
Robert Stephenson years afterwards declared that it was one
or the greatest triumphs in the art of bridge-building ; and at
that time it undoubtedly was.   No one had ventured upon
such a span, and in iron too ! So cast-iron bridges threw stone
and timber into the shade.

Telford, an eminently practical man, soon adopted the new
method.   He was of course aware of the progress of the Wear
bridge and about the same time he designed a new iron bridge
to cross the Severn ; " a flat arch," was the main feature of this
still existing bridge at Buildwas.

It would be tedious as it is unnecessary to continue the
details of the many other works connected with the name of
the great road-maker.   No doubt the incidents connected
with their initiation and construction would be in some ways
interesting though of a sameness.   We will however pass on
to the development of the highways in other parts of the kingdom in which Telfold assisted.   In the appendices to his
" Life " are many reports and descriptions of Scotland, for
whose benefit he was so usefully employed in the early days
of the nineteenth century.   Let us therefore turn to the North

It will be, then, necessary to pass over many of the public
acts of Telford, as a constructor of canals and bridges, to reach

THE WEAR BRIDGE, SUNDERLAND.

the period in which he was deputed by the Government to undertake the repair of the roads in Scotland. We have already seen in what manner these rough highways had been cut by General Wade. In 1802 the condition of the Highland roads attracted the attention of the Government, and a report was printed. The highways were in a very defective state, " abounding in holes," the ascents steep and their liability to be injured by torrents very great. So, in consequence of the difficulty of communication, cultivation was almost entirely confined to narrow strips of land near the sea coast. Bridges were almost unknown, ferries were not frequent, and yet they were the only means of crossing the dangerous streams and rivers. The people lay idle, miserable, slothful, did nothing save distil illicit spirit, and made no exports. When the roads were constructed by Parliament a tremendous change set in immediately. Produce was exported, vehicles abounded; people, lately almost savages, became civilised ; the pig and the cattle were " treated to a separate table " ; the dunghill was put *outside* the cottage ; the tartan tatters gave place to the produce of Huddersfield and Manchester, Gaelic to English, and few people could be found who could not read or write.

Until the Marquis of Stafford began his improvements the plough was unknown in Sutherlandshire. The ruinous bridges, one of which caused the fall of the mail coach and horses into the river, the death of the coachman and one passenger, and the maiming of many others, remained unmended, and the means of communication were very limited everywhere.

Even this accident did not compel the renewal of the bridge till the end of the century. The fallen side was merely fenced off, and carriages, mails and passengers were compelled to pass upon the only remaining portion in single file ! All these inconveniences and dangers, all the benefits enumerated above, were respectively avoided and conferred by the road-engineer.

Telford gives us specifications and details of his work in Scotland, including the great Caledonian Canal, which we shall refer to in another section. He constructed road after road, built bridge after bridge.   The local nobility and landowners assisted, and so within a comparatively short space of time the the Tay, the Spey, the Severn, and many other rivers were bridged in an elegant and substantial manner, greatly to the advantage of the country.

In addition to these labours, our engineer and his assistants opened up by road the trade and commerce of Scotland. Wherever a road seemed desirable Telford made it, connecting his main lines, his bridges, his canals, and his harbours with each other and with the interior of the country.   These roads we are informed necessitated the building of 1,200 bridges of various kinds.   Thus rapidly numerous bridges were built, streams and ferries conquered, posting increased, travellers multiplied ; the judges and counsel, hitherto compelled to ride, and run tremendous risks, were enabled to drive in comfort from place to place.   The wand of a magician could scarce have made such a change ; the country was opened up, its natural resources and natural features were appreciated, and Scotland stood where she had never stood before.

No longer was the horseman in danger of being swept down the current at the ford, no longer was the leaking boat impelled across the stream with horses swimming astern !   The rate of progress of a mile an hour was changed to the swiftness of the Flying Coach, and " Edinburgh could then be reached from Glasgow in *ten hours* ! " a triumph in those days.

Of course this increased pace was due to the improvement in the roads, owing to the employment of the methods of road making of Macadam and Telford.

CONWAY SUSPENSION BRIDGE.

CHAPTER IV.

MACADAM AND HIS SYSTEM.—THE HOLYHEAD ROAD AND IRISH
TRAFFIC.—OLD SUSPENSION BRIDGES.—THE MENAI BRIDGE
AND ITS STORY.—THE BROOKLYN BRIDGE.

R. MACADAM—Loudon Macadam—reached
England in 1783 from America, and became a
surveyor in Bristol. His son James was associated
with him in the working and making of the highways upon which, in Scotland and elsewhere, Thomas Telford
was also concentrating his mind.

The principle of Macadam was a solid road of broken stone
consolidated by pressure; not bound by other material, but
composed of broken stone of no more than a certain specified
size, and weight of six ounces. Stone-breaking then became
an occupation, as it still remains a punishment in our work-
houses for those convicted of the crime of poverty.

James, afterwards Sir James, Macadam thoroughly carried out these principles, and continued his father's methods. At first sight it may appear curious that a coach-proprietor should prefer a hard flint or granite road to a soft, or chalky, or gravelled road. But the worth of Macadam's practice became evident when the regular-sized stones became consolidated by pressure and united into a hard smooth surface not easily disturbed by horses' feet which cut up the old roads terribly. Macadam's method drained and hardened the once heavy, muddy, rut-full highways. The road mender was in evidence, the heaps of broken granite were always handy by the roadside, and though Macadam died poor, he died "at least an honest man," as he declared.

Telford had the advantage in his method, which he introduced first in Scotland, where he seems to have anticipated Macadam in the limitation of the size and weight of the stones used. But however that may have been, the mode and practice thus indicated are known as Macadamization, and the highways are Macadamized, not Telfordized !

It was in 1815 that Macadam devoted himself to making roads, but Telford's fame gained in Scotland ensured him employment in England. His bridges and canals were already in evidence, and the Government began to realise the advantages of good roads for traffic, now that the selection of horses for the mails did not depend upon Hobson, whose "choice" is still a proverb, nor on the enterprise of the manager of the Bath theatre, Palmer, who had so greatly improved the coach.

The enterprise which is most generally connected with Telford's name, and the one which we may select as typical of road improvement, was the construction of the Holyhead road and the Menai Bridge. Nothing can show the state of things more clearly than the evidence laid before the Parliamentary Committees in the reports made by them upon the Holyhead road and harbour. We, in these days of railroads and swift steamers, cannot grasp the extent to which inconvenience reached in the early years of the century. Readers of Charles Lever's novels can picture the misery of the passage across the Irish sea from Dublin to Holyhead, before Kingstown had received that name, and Dunleary was a fishing village, *sans* piers and jetties, harbour or lighthouse.

MENAI SUSPENSION BRIDGE.

The state of affairs is described in books of travel and diaries. Miss Burney gives us many glimpses of coaching life and customs, and to Mr. Roberts, for his glimpses of social life early in the century, readers may turn for information.

In those days posting was expensive. From London to

Holyhead the journey occupied four days, and cost perhaps seven pounds. The expense, when all the members of a family travelled, accompanied by men-servants on horseback, and followed by baggage-coaches, became very great on the highway; to it the transit in the ferries was added—a considerable cost. Thus communication between England, especially the south of England, and Ireland was at once limited and expensive.

Telford, at the instigation of the Government, set about to remedy the then existing condition of affairs. At the time of the Union there were three ways of reaching Ireland, viz. *via* Milford and Waterford, by Holyhead and Dublin, and by Portpatrick and Donaghadee. Passengers were landed at the North Wall in Dublin, and as the vessels were only sailing-packets with limited accommodation, the miseries of the transit can be imagined. Steam had not then commenced to ride the waves, and a tossing for several hours was the least of the inconveniences. Sometimes many days were consumed in the attempt to cross, and when Wales was at length reached, the rugged, rocky, and stony road, and the jolting conveyance received the miserable voyager, and upset him once more in every sense.

Leaving Holyhead the so-called road—a narrow track—crossed Anglesea to the strait, where "a troublesome and dangerous ferry" had to be crossed. This was a tidal ferry, and at certain times the greatest care had to be exercised in the passage; accidents were numerous nevertheless.

Then, supposing this Rubicon were safely passed, the transit was continued upon "a road, generally speaking, narrow, steep, and unprotected by parapets"; and even after reaching smoother ground, the mail-coach road was in a very imperfect state, for trees had to be lopped in places to clear a way for the loaded coach.

So, naturally, "business was impeded," and the Irish Members had considerable difficulty in reaching their constituents. In Rickman's "Life of Telford," we read how the improvements were begun by Mr. Rennie, who surveyed the landing-places of the "packets" and decided in favour of Holyhead and the long promontory of Howth, which juts into the channel north of the bay and north-east of Dublin.

Travelling over such roads presented many curious features : the "arbitrary bashaw," the ferryman, the contemplation of a hanging highwayman. "The jingling of bells" warned the approaching traveller in narrow places ; and travellers meeting at cross-roads would halt and enquire for the news ! Carriers' teams sometimes extended to thirteen horses, and although an

HOLYHEAD LIGHTHOUSE.

Act was passed limiting them to nine, the carriers bribed the informers, and went on as before ! One member of the fraternity declared that " roads had no object save for wagon-driving. He required five foot in a lane, and all the others might go to the d——! The gentry ought to stop at home, and not run gossiping up and down the country."

The cost of a family party on the journey from Dublin to London is given at £114 3s. 6d. ; and when we think that a bottle of brandy for the transit of the channel cost a guinea, the total expense will not be wondered at. The Bangor Ferry

is put at £1 10*s.*[1]   In Mr. Roberts' work we also see the
origin of the term " lionise."   In the early days the fashion-
able things to show one's country friends were the lions in the
Tower.   Every traveller to London was taken thither, hence
the term " lionise."   But this is a disgression.

To remedy the many ills of the road Telford laboured
hard, and yet nothing was done until about 1815, when the
often occurring accidents, the upsets from racing, and breaks-
down during the forty-one hours' drive to Holyhead, compelled attention.   Sir H. Parnell went in vigorously for the
Reform of the Road, and carried his point.   By the year 1820
the Holyhead road was in good going order, well levelled and
"finished off" in Telford's best manner.   But admirable as
were and are his Welsh roads, there was a missing link which
required to be forged.   Two ferries interrupted the traffic, and
so bridges across the Menai and the Conway estuaries became
prime necessities.

These celebrated bridges, which are rivalled by the
Britannia and Conway tubular viaducts, are now accepted as
commonplace structures, mere connecting links in the chain
of communication.   But some fifty years ago or so, this
indifference had not set in ; the bridges, whose existence is
almost ignored, were then the general topic of conversation,
and even as late as 1849 excited much curiosity.   The account
of the erection of the Menai Bridge, will, we think, prove
interesting, after such a lapse of time, to young readers who
mayhap have given no thought to the peculiar circumstances
of its construction and of its novelty at the time.

It is true that Runcorn Bridge had been designed by
Telford upon the suspension principle, but it had not been
erected.   The engineer had made many experiments.   He
knew that the suspension bridge had been in use in India
for centuries.   It is, perhaps, one of the oldest forms of
bridge, which in the Himalaya is known as a Jhula, and we may

<hr/>

[1] " Social Life in the Southern Counties of England."

here give some description of it before proceeding to relate the triumph of suspension as exemplified in the Menai Bridge.

There are Jhulas of sorts, the most common being those constructed of grass or fibre-rope. Some consist merely of a rope and a kind of basket, in which the delighted traveller is swung like a sheep across the torrent which roars and rushes in the middle of the passage, within an easy measurable

BRITANNIA TUBULAR RAILWAY BRIDGE.

distance of his cradle which may be his grave. This kind of bridge is really only a rope on which the natives cross, hanging by hands and feet in a loop. But for less favoured individuals there is a kind of basket, which, if the rope slopes, runs one way by its own momentum. There are guy-ropes attached, by which it is pulled back.

On one side of the stream is a stout two-pronged fork of wood. Around and through the prongs the rope is placed and fixed on shore something after the manner of the rope on which an acrobat performs. The opposite end is carried over and fixed in the adjacent rocks. The rope, well greased, supports the basket, and it travels backwards and forwards when required.

Should your guides or native companions find any difficulty in obtaining their demands, you may find yourself left high and partially dry in the centre of the rope, until their requests are complied with.

There is another Jhula, constructed of some worsted ropes, on which runs a block of wood, which is drawn back and forth over the river. The passenger is secured by certain loops which are passed round his body, and he is hauled across like a sack.

Bridges on the suspension principle, made of grass or fibre ropes, are very picturesque but rather dangerous, for as the repairs are generally postponed till some few passengers have lost their lives, the risk in passing is considerable. These rope-bridges sway and "sag" within a few feet of the stream at times. The ropes are about the size of a cable, and made of the coarse grass, which is plentiful, in three strands. These ropes support the "flooring," which is really a kind of ladder wattled with twigs, and fixed to the guide-ropes by pendent cords. If the banks are high the passage is rather a curious and a nervous business; if the banks are low the rope hangs just above the water, and a freshet is unpleasant.

Perchance you may find yourself on the hither bank of a roaring, plunging torrent, dashing down in its headlong course amid stupendous rocks and boulders. The Jhula, or twig-bridge, is described in several works of Indian travel and exploration.

The bridge in question is formed of twisted twigs of the birch-tree made into ropes of considerable thickness, and hung about five feet apart. Between them runs another rope a few feet lower than the side ropes, "being connected with the upper ropes by mere slender cords also usually of birch twigs twisted together," and occurring at intervals of about five feet. The mode of progression must be very unpleasant, and is thus described by Mr. Andrew Wilson :—

"The unpleasantness of a Jhula is that the passenger has no proper hold of the upper ropes, which are too thick and rough to be grasped by the hand, and that at the extremities they are so far apart that it is difficult to have any hold of both at the same time, while the danger is increased by the bend or hang of the Jhula, which is much lower in the middle than at the ends. . . .

"The traveller must stoop as he proceeds, and his progress must be continuous : it is unwise to pause on this swaying bridge, and dangerous to look down, as then the frail structure seems to be racing up-stream at a tremendous pace. Besides the danger from giddiness there are the chances, and not very unlikely chances, of the rope giving way. These Jhulas," says Mr. Wilson, "are not usually repaired till some one has fallen through and announced the rotten condition of its materials." A passage some hundred feet on such a bridge is unpleasant, to say the least.

Of course, Telford had no intention to build a bridge of such risky sort. His mind was set upon a suspension bridge of iron ; of cast iron " chains," which would in mid-air, depending from piers, support a roadway. His Runcorn Bridge had not been proceeded with, but he was none the less confident of his project. Many eminent persons agreed with him ; but, on the other hand, there were many detractors who opposed a suspended bridge on the plea that it would interfere with the navigation !

But when the idea took tangible shape it interested every one. People remembered the prophecy of the mysterious "Robin Dhu," whose characteristically " dark " sayings had included a prediction that a bridge would span the strait near the Bangor Ferry. The right of ferry had been granted by Queen Elizabeth to John Williams, an ancestor of Lady Erskine, in 1594, for a rent of £3 6s. 8d.

The first step taken in the direction of the bridge was the

E

levelling of the rock on which the western main pier was to
stand. This rock rejoices in the name Ynys-y-moch, and the
prediction just referred to foretold that a bridge would one
day be carried across the Menai Straits there. This beginning
of the suspension bridge dates from May, 1819.

When, despite opposition, powers had been obtained from
Parliament, the navigation was entirely stopped, and a causeway was built across the strait to facilitate communication.
The first stone of the bridge was laid by Mr. Provis, resident
engineer, at ten a.m. on the 10th of August, 1819, without
any ceremony.

·Progress was made but slowly, and the contractors were
superseded by others. Meantime negotiations were proceeding
for the purchase of the ferry, and, by arbitration, the sum of
£26,394 was paid to the Erskine family—thirty years' purchase of the income.

For a long time severe weather interfered with the work, and
it seemed as if the "elements" had a special spite against the
bridge. The barges were wrecked, the pier and jetty swept
away in places, and so, notwithstanding all efforts, progress
continued very slow. But perseverance succeeded. The piers
and arches rose; the rock-tunnels for the chains were driven;
the iron for the chains was tested, and when dipped in oil and
warmed to a moderate heat, was found to be secure from
the effects of the atmosphere.

Notwithstanding the weather the first iron work was fixed in
March, 1823, but the time for the completion of the bridge
had to be extended until July, 1825. So the work went on
till April in that year, when it was intended to have put up
the first chain. An accident delayed it till next day, the 26th,
when a number of the *élite* and the populace assembled to see
the operation. The chains had only been carried from their
rocky fastnesses as far as the summits of the suspension piers.
The operations to be performed now were the joining of the
chains, and then the raising of them.

It was a pretty sight. The weather was fine, and many

pleasure boats dotted the strait. Flags waved and fluttered, carriages drove up on either side. Precisely at half-past two p.m. the immense raft, which sustained the end to be brought across to the Anglesea side, quitted her moorings, carrying the chain—the bridegroom, as it were, who was to be united in indissoluble bands to the Anglesea chain. The raft was swept out by the tide to the centre of the bridge and made fast, thus the raft extended from one side to the centre of the bridge underneath.

The opposite end of the chain was then hauled in, and the end links of each were united by a screw pin on the raft. Capstans, manned by many willing workmen, then began to wind up the tackle to a lively tune. Up came the hawsers, up came the heavier united chains now one and indivisible. The tide now turned ; the raft was swung back on her anchors and let go. When it was seen that the mighty chain hung gracefully in the water like an inverted bow, cheers rang out loudly. The hauling process continued, and at five o'clock the last rivets had been driven, the first chain hung completed 120 feet above the Strait.

Tremendous cheers greeted the result. Mr. Telford was standing on the pier to witness the triumph of his brain. And scarce had the congratulations ceased, when a workman got astride of the chain and worked himself across it from shore to shore. Two others followed, one of whom actually walked part of the distance on the nine-inch chain !

A second chain was carried over on the 28th April, and when it had been securely fixed a workman sat upon the upper chain resting his feet on the lower, and there made a pair of shoes in two hours, in the presence of the spectators. On Saturday, 9th July, the sixteenth and last chain was fixed ; and on the 21st, the navigation of the Strait was re-opened by the steam packet *Saint David.*

The bridge was then nearly ready, and the Ferry would shortly be a relic of the past. On the day before the opening of the bridge notice was sent to the Ferry-house that after the

mail-coach had passed the bridge the Ferry must cease to ply.
On Sunday, 29th July, 1826, the structure and the works were
examined by Telford and Sir H. Parnell.   The lamps were
lighted, but as the evening was stormy the engineers did not
remain to cross in the coach.   "Everything was put in readiness," says Mr. Provis, and, "aware that the coach would be
well filled before it reached the bridge, I went with my brother
and met the mail before it reached Bangor.

"Taking my seat by the coachman, he and the guard were
then informed that, instead of stopping at Bangor Ferry as
usual, they were to drive over the Menai Bridge, where the
horses would meet and take them forward.   There were four
passengers inside the mail.   When opposite the Ferry inn
Mr. Akers and several others joined us, and on stopping for
a moment at the end of the bridge the mail was instantly
crowded by as many as could find a place to hang to.   Thus
loaded the coach passed over amid the whistling and howling
of the storm."

The down mail also crossed at half-past three a.m., and
next day at daylight, flags denoted that the suspension bridge
was open to the public.   Severe gales tried the structure,
which remained firm, and in this manner the crossing of the
Menai Strait was accomplished.   Southey commemorated
this in his lines to the engineer at Banvie on the Caledonian
Canal:

> "Telford it was, by whose presiding mind
> The whole great work was planned and perfected—
> Telford, who o'er the vale of Cambrian Dee,
> Aloft in air, at giddy height upborne,
> Carried his navigable road, and hung
> High o'er Menai's Straits the bending bridge."

The total length of the bridge is 1710 feet; the space
between the suspension piers is 579 feet.   The iron employed
weighs more than 2,000 tons, and the cost of the structure,
including approaches, was £120,000.

The Conway Suspension Bridge was subsequently erected on the same principles as that over the strait, and both remain monuments of Telford's genius.

BROOKLYN BRIDGE, NEW YORK.

It is not necessary to pursue this road farther. The high-ways were made, and many magnificent bridges have been constructed during the last twenty years in all parts of the world for roads and railways. Some of the latter we may touch upon later. But the great suspension bridge at Brooklyn must not be ignored. It is the only rival of our Forth Bridge, though the latter is on the cantilever system, which, as applied to wooden structures, is as old as the Chinese.

Let us look now at the Brooklyn Bridge, in comparison with the Menai Bridge. Its dimensions are interesting, and are as follow :—

The East River Bridge is something over a mile in length that is to say, 5,989 feet. The centre span is 1,595 feet 6 inches, with two side spans 930 feet each. The roadway is raised in the centre of the span, where it soars 136 feet above the water. The structure is of steel, a suspension bridge, and crosses the river on a slight slant, following the ferry route slightly down stream. The gentle slope of the bridge to the centre is very graceful; the rise is fifteen feet from the sides.

The story of the construction of this really magnificent bridge would occupy too much space in the telling. The company was organized in 1867, and the bridge was opened in 1883. By incessant labour, and caisson work in compressed air, the towers and anchorages were completed in 1876, but an immense empty space lay between them. To the uninitiated it was a puzzle how this enormous gulf was to be spanned. But the engineers were prepared to answer the question.

They provided a rope of length sufficient to stretch across the stream over the tops of the towers. The barge containing this wire cable was towed across the river, and the rope hauled up the face of the New York tower, and made fast. Another cable was then carried over and united in an endless coil with the first. This rope was passed over drums, and when steam power was applied the rope travelled over and above the stream continuously.

This connection made, it became necessary to utilise it. The mechanics had to work suspended in mid-air, but some one must " bell the cat," and set an example; no workman would venture, yet Mr. Farrington did. He put together a swinging seat, and slinging this board to the travelling rope, was run across the intervening space amid the cheers and shouts of an immense concourse of people and the roar of cannon. The rope " went hissing and undulating like a flying-serpent through the air," writes an eye-witness. Away went the whirring rope, invisible, or like a spider's thread, to the spectators, "bending and swaying with the human weight that rode its cantering waves."

The enterprising engineer crossed in safety in twenty-two minutes. After that day, 25th August, 1876, the progress made with the suspension spans was rapid. But in June, 1878, a terrible accident happened—one of the strands of the cable got loose. The result can be hardly imagined: we may perhaps picture the sweeping steel rope rushing with all its ponderous weight across the tower platforms, cutting down and sweeping into the river a number of mechanics.

The rope made one mighty leap across and over streets and houses ere it plunged into the river. Providentially it did not strike any of the numerous ferries or other craft in its descent, or the loss of life must have been terrible. This was the only serious accident, and in the summer of 1883 the Brooklyn Suspension Bridge was opened.

A magnificent prospect can be obtained from the roadways, of which there are five. The outer ones are reserved for ordinary carriage traffic, the centre one is for pedestrians, and the other pair between these are for the tramcars which, worked by a rope, run steadily and rapidly over the bridge. Till the East River was thus spanned, the Niagara Suspension Bridge was the largest, and it is still reckoned as one of the most remarkable of bridges. It has a span of 822 feet, and is 245 feet above the water. There is a more modern railway viaduct of the cantilever pattern near it. Of this we may treat later, if space permit.

We will now leave our bridges and carry our readers across the Channel for a glimpse at some other roads, and conclude with a short account of the great carman, Charles Bianconi.

HORSE-LITTER, TEMP. RICHARD II.

CHAPTER V.

FOREIGN HIGHWAYS.—TRAVELLING AND CONVEYANCES.—THE STORY OF CHARLES BIANCONI AND HIS CARS.

O far we have limited our excursions into the realms of highways to England and Scotland, where we have traced the development of the road from the footpath or the horse track down to the Macadamized thoroughfares of the present day. We have endeavoured to describe the gradual rise of engineering skill in the department of bridges, and have selected a few for illustration, as instances and types of the many hundreds of similar structures in stone and iron which abound in the United Kingdom and in the world beyond it. But there are numerous other roads whose praises should be sung. We may here only enumerate them, for the railroad has pushed them into the background more completely than in England.

To mention only a few. The grand St. Gothard road was dominated by the railway, the scene of many conflicts between French and Russian, and Austrian, in those days when

Switzerland was the battle ground not the playground of Europe, when the, now disappeared, Devil's Bridge was blown up, and the opposing Russians, binding planks with their scarves, crossed the chasm and carried the position. To this mountain road succeeded the carriage road, a splendid piece of engineering, winding up to Hospenthal beside the rushing Reuss, whose waters once ran red beneath its old-time bridges

TRAVELLING COACH, SEVENTEENTH CENTURY.

All is changed now. The highways in the cantons are not frequented, but grim testimony to the sanguinary scenes of 1799 and later years remains. Here is the Lake of the Dead, on the summit of the Grimsel, where Austrians and Frenchmen filled the waters with their killed and wounded. There is the Splugen and the gulf of Cardinell along which in 1800 Brune led the French divisions into Italy, and into which the little drummer was swept by an avalanche. There his numbing fingers continued to beat his drum—down in the gorge its sound was heard, but no help could come, and "death stilled his icy fingers."

Again, the famous and fatal retreat of Suwarow through the
Muotta Thal makes us shudder. The Russian general, forced
to give way, led his army from Altdorf, and some idea of the
width of road may be gained when we mention that when the
vanguard had reached Muotta, eight miles from Schwyz, the
rear guard was at Altdorf in the St. Gothard. In this Thal the
French encountered him and the awful retreat began, after
obstinate fighting, by the Pragel, and so to Glarus. Then he
was compelled to pass the Panix in the Grisons, where his
soldiers perished by hundreds daily, marching single file in the
snow to death. In this retreat Suwarow lost 8,000 men.

These are but a few of the scenes which the highways and
by-ways of the Alps have witnessed. The restless Napoleon
dragging his cannon over the St. Bernard pass, or impatiently
demanding "when can the artillery pass over the Simplon," bring
to our minds great engineering feats of ancient and modern
times. The St. Bernard, used a hundred years B.C. by the
Romans, was crossed by Napoleon in 1800, A.D., and is the
scene of "Excelsior." The Emperor's difficulties in his passage
of the Bernard determined him on finding another road. In
1801 five thousand men were set to work on the road across the
Simplon, and completed their work in five summers. The old
path, named Sempronius after the consul, was thus converted
into a splendid carriage road. It costs eighteen millions of
francs, and there are several tunnels and galleries, and 611
bridges on the route. A splendid road soon to be superseded
by the all conquering railroad. Well may the Swiss have put
up the inscription in the Splugen Route: "The way is now
open for friend and foes. Swiss be vigilant !"

We must pass on from these reminiscences. Many curious
and romantic incidents still cling to the barren rocks and steep
horse-tracks of the mountains, but we must not examine them.
Within our own islands the road can furnish us sufficient
material in the adventures of coach and car and chaise. The
post-boy and the postillion have departed. The coachman
and the guard are still with us in the playing at old time

coaches which run so rapidly and so pleasantly from the Métropole or the classic "Cellar." We have our coachman still with us, a sprig of nobility mayhap—a nobleman may be, or a regular "whip." But where is the old coachman, what was he like? Read "Pickwick" and look at Mr. Weller, senior, addressed affectionately by his son as "my old Proosian Blue," or "a old image," according to temper and circumstances.

Dickens painted no fancy portrait. Here is a contemporary sketch, "a tall, corpulent man," whose ruddy weather-beaten visage was partly shaded by a broad-brimmed, low-crowned hat; and a fat double chin was encased in the ample folds of a blue spotted shawl. A long striped waistcoat, approaching his knees, was buttoned closely over a portly body; and a pair of drab breeches, with fawn-coloured ribands, dangling in graceful neligence at his knees, adorned a couple of tubby-looking legs. The coat, which afforded protection not only to his ample shoulders but to his heels, was of faded brown, and highly-polished lace up shoes completed the attire.

In those days people wore "great" coats indeed. Passengers, guards and coachmen alike wore garments with capes and voluminous skirts. Many of these "whips" were incompetent. Why? "Simply," writes an authority in 1828, "because the *excellence of the road* annihilated the breed!" The excellence of the highway, the result of the efforts of Telford and his contemporaries, had caused a corresponding degeneration in the drivers of coaches.

This is a curious fact, but perfectly authentic. Out of forty-five coachmen the writer already quoted can only approve eight capable of that art of "hitting 'em and holding 'em," which is the true secret of driving. This was apropos of the Brighton Road.

Those who may require an almost veritable "photograph" of the old coaches in connection with which so much of the romance of the highway, after the roads were fairly well made too, can read the "Tale of Two Cities." In the early part of this novel Dickens gives generations the result of his observation and inquiry. Look at the coach, the old mail, with its damp and dirty straw, its disagreeable smell, and its obscurity. Travelling in those days (1775) must have been a very unsociable business in the mail, when "travellers were very shy of being confidential, for every one on the road might be a robber or in league with robbers." Coaches then set out with a perfect arsenal of firearms. An arm chest was as much an item of its equipment as was a tool box; in the former were "a loaded blunderbuss on the top of six or eight loaded horse-pistols, deposited upon a substratum of cutlass." And, when the mounted messenger overtook the mail on the summit of Shooter's Hill, you will remember how the guard called upon the passengers in the king's name to stand by him and repulse the expected highwayman. Pleasant times those on the great but muddy roads out of London, when, if any one of the passengers walking up the hill in the light of the coach lamps had suggested to another to walk on a little into the darkness, he would have run the risk of being shot for a highwayman.

We can scarcely conclude our notice of the stage-coach on the highway without glancing at some other of the quaint vehicles which in times past adorned our paths and roads. We have already mentioned the pack horse; long cavalcades with burdens used to cross the hills and make their way to London by the same undulating track, the same thoroughfare as is now called a "horse path" in Cumberland or Switzerland. By means of these panniered animals smuggling was extensively carried on. Some young people when reading tales of smugglers may have thought how stupid was the coastguard not to suspect a pack of horses or mules as likely bearers of ill-gotten property. But such a reader should remember that such modes of conveyance were in early days most usual, and in comparatively late years not uncommon; and therefore there was nothing more unusual in a convoy of pack animals than there is to day in a string of four-wheeled cabs, or hansoms.

Of the vehicles in olden time perhaps the earliest is the vehiclicote or wheeled bed, which was introduced into England in the reign of Richard the Second. This kind of litter on wheels was used by invalids and others who were temporarily or naturally incapacitated from riding. It was something like a hammock on wheels, slung from two perpendiculars. The horse-litter had been its predecessor, and this horse-litter was a kind of palanquin slung on poles or shafts which were supported by horses ridden by men or lads. The sedan chair is but a modern imitation of the litter or "chair." The reason for the name is uncertain, but it did not emanate from Sedan town. Mr. Sala suggests that it was so named from the peculiar cloth called "sedan," with which these conveyances and gondolas were lined.

The private coach or carriage did not come into vogue till Elizabeth's time. Readers are familiar with the figure of the "Virgin Queen" seated on horseback, and this is conclusive proof that carriages or "coaches" were not used much in her reign. But even her coach was more like a springless cart, and by no means comfortable. A coal-cart of the present time is not the most easy of vehicles, but a springless cart on a rough country road in the sixteenth century must have been terrible.

The Queen's coach set the fashion, however. People obtained the like, but found them as uneasy as the head that wears a crown.

Besides, the narrow streets already mentioned precluded the use of carriages, and it was this fashion which tempted people to live out west where the country was open and pleasant. No one could have been pleased with the jolting and the frequent upsets in the muddy, rutty, and stony ways.

However, in that reign or in the next, some ingenious person invented a kind of strap suspender which took off much of the jolting of the carriage, and we find coaches *temp.* Charles I. hung in this way by straps. But in the time of the merry monarch steel springs came into fashion, and in the immortal

" Diary " we finds Pepys noting the occasion, in 1665, on which he went to see some contrivances for making coaches go easy, and he and his friends " rid " in the long-springed vehicle.

One of the most romantically associated vehicles of any period was the post-chaise. In it many adventurous couples hurried off from paternal or avuncular wrath, and in them pursuit hot and speedy followed. When Mr. Tupman was so deceived by Jingle, the adventurer and the lady were pursued by Mr. Wardle and Mr. Pickwick in the post-chaise and four from the Blue Lion at Muggleton. Post boys were then common objects of the highway, old and wizened little men who had mayhap ridden " post-haste " in their youth with his Majesty's mails.

In these not very comfortable vehicles our ancestors travelled. Then came the brougham, the tilbury, the stan-hope, names derived from their inventors or introducers, at least this is the case with the two former " traps." The stage coach body was in shape very like a single compartment of a railway carriage or " coach " at the present time. A high box in front and a large basket behind, between the hind wheels, were features of the " stage." Travellers rode in the " basket " at the back where they got jolted pretty freely.

The populace travelled, when they did travel, by wagon or by canal boat. The writer can remember the " fly-boats " on the canals from Dublin carrying passengers. The boats were drawn by three horses tandem, and travelled at a quick trotting pace some twelve miles an hour. These boats with cargo-boats and jaunting cars were the general means of transport. The " inside " car was used in wet weather. It was simply a black box, with seats, perched on wheels, with four small and one large window, the last over the door. Many curious tales are related of the " inside car," and when some worthy guest's " inside " was mentioned, smiles would go round amongst the English strangers present who did not understand what was meant, and the quaint announcement of the approach of the

vehicle from the yard was provocative of alarm and amusement.

Apropos of "jaunting cars." The national vehicle of the
" most distressful country." The career of Charles Bianconi
deserves some mention in connection with Roads and
Romance. We well remember his "long cars" which, made
wagonette fashion, painted red, carried mails and passengers
through many parts of Ireland. We have read somewhere
that Bianconi "invented" the jaunting car. This we believe
is a mistake. The Irish jaunting car is of older date than
Bianconi's arrival in Ireland, whose communications he did so
much to open up.
Mrs. O'Connell tells
us a great deal about
Bianconi, and though
he was not an en-
gineer, and performed
no engineering feats,
he was so intimately
connected with the
development of the
roads and consequent
prosperity of Ireland,
that he must be sketched here.

JAUNTING CAR.

Bianconi came from Italy with a person named Faroni, by
whom he was employed in selling prints, mounted pictures,
with which Charles and his associates "travelled." They
began in Dublin in 1802, and Bianconi afterwards passed
through the South of Ireland, taking notes in his mind and
silver in his pocket, until his apprenticeship to Faroni had
expired. Declining to return to Como, he started business on
his own account, and being unable to afford conveyance had
to carry his heavy pedlar's pack.

This burden set him thinking how he could compass conveyance, but the idea did not form itself into practice until a
few years after when, living and trading as a carver and gilder

in Clonmel, the idea again occurred to him to utilize the "outside" cars for a regular service between that town and Cahir. Bianconi used to declare that the car scheme "grew out of his back"; and he certainly owed its initiation to his heavy burden of pictures.

People in Ireland were too poor to ride in the coaches or mail cars; they did not often ride on horseback for the same reason, and walking with heavy loads was distressing, as Bianconi knew very well. His first trial of the plan he had contrived was made in the year 1815, after the pride of Bonaparte had been humbled, and peace was settling down in Europe. The first Bianconi Car started from Clonmel to Cahir, twelve miles, on the 5th of July, 1815, the fare was eighteenpence; but no one patronised it! Not a single passenger tried the car; the peasantry could walk gratis why pay for a "kyar"?

The scheme would doubtless have failed if the originator of it had not been a man of resource. He soon started an opposition to himself, and thus stimulated the sporting tastes of the Irish. The people became interested immediately, the cars raced along the road, and at times took up passengers for nothing! When Paddy found that he could get a "set down" gratis he took it, and thus became accustomed to riding on the car. Naturally somewhat indolent, the men clung to the "jaunt," and agreed to pay for it. Equally naturally the people espoused the causes of the rivals, and some travelled on one, some on the other car. Thus money came in fast, until one inauspicious (?) day the opposition cattle "broke down," and Bianconi was left master of the road!

By this time the residents in the district were "spoiling" for a drive. They must enjoy the accustomed luxury, and besides, they found that driving was in the end the cheapest course for business. The car was fully patronised, another was started, and by degrees others, so that very soon Bianconi had cars running through several districts.

Then the "carver and gilder" set up a car manufactory, and abandoned his former business. His misfortunes helped him. Pelted by the populace for assisting a non-national candidate at the Waterford election, when two cars were destroyed by the adherents of the people's representative, Bianconi was released from his engagement to carry Beresford's voters to the poll. He then was engaged on the other side, which he assisted to win the election, and was paid a thousand pounds for car-hire! Thenceforward his success was assured.

From that time (1826) Bianconi prospered. He opened up communications in many districts, west, south, and north-west. The tourist cars were improved, and all the conveyances ran in connection with each other, so correspondence was ensured. "They virtually opened up seven-tenths of Ireland to civilization and commerce . . . and the magnificent scenery of Ireland to tourists."[1]

No one interfered with Bianconi. No Fenian or highway-man ever molested the "kyar." No Whiteboy challenged it. . . . The energetic owner got the mail contract, and im-proved the service, but his coach was eventually a loss, owing to the withdrawal of the mail contracts; but Bianconi ran the coach all the same.

By insisting on punctuality, sobriety, and civility, by pro-viding good horses and good men, every one of which and of whom Bianconi knew by name, this immigrant benefited Ireland more than any Acts of Parliament. In his evidence before the committee on the postal service, and in Mr. Smiles' volume already quoted, will be found many more particulars of Bianconi's later years. The writer can distinctly remember the time when Bianconi's cars were as "familiar in our mouths as household words," and the long Killarney "kyars" were filled with tourists. Pictures of these may perhaps be found in contemporary "tours," such as Mr. and Mrs. Hall's work, and others, in 1840–50.

---

[1] "Invention and Industry."

Charles Bianconi died in 1875, aged 89, greatly regretted. He superintended his business to the last, and in his seventieth year "he was still a man in his prime; and he might be seen at Clomnel helping at busy times to load the cars." We will conclude this notice with the eulogy pronounced upon him at the meeting of the British Association at Cork :—" Although Ireland might claim Bianconi as a citizen, yet the Italians should ever with pride hail him as a countryman, whose industry and virtue reflected honour on the country of the birth. He was a blessing to Ireland indeed."

CHAPTER VI.

TURNPIKES AND TROUBLE.—REBECCA AND HER DAUGHTERS. —THE RIOTS SUPPRESSED.

E have seen how the highways were improved, and how the Turnpike Acts were passed; but there is a phase in our romance in connection with the public roads which should not be passed over—we mean the revolt against the turnpikes.

The highway was not necessarily a turnpike road; and we know that some roads only were called "turnpike," and the toll-bars "gates." These tolls were farmed out by various people who appointed men to collect them. Such men, secure in their huts, defied the traveller, and few, save the daring highwaymen, ventured to "clear the gate" by any other mode than by payment.

These "pikes," as with the British love of abbreviation the toll gates were called, caused considerable annoyance in various parts of England. More than a hundred years ago, an outbreak occurred against the toll-gates which had been erected in accordance with an Act passed, for repairing the roads ten miles round Bristol, in 1749. The Kingswood colliers murmured loudly, and after a couple of weeks Ashton turnpike was destroyed during the night, and another barrier shared the same fate a little while after.

The turnpike Commissioners did what they could under the circumstances. They offered a substantial reward for the apprehension of the offenders, and proceeded to re-erect the gates. These efforts were of no avail. When the Bilton turnpike, which had been "hoist" with gunpowder, was repaired, the rioters cut it down.

Three men were immediately arrested and put in prison; but the friends and allies of the pike-slayers determined to have them out. With this view an immense mob assembled on the 1st of August, and in full daylight marched into Bristol in order, with music and banners, and "commanded" by men on horseback. All were armed with offensive weapons, and the city seemed doomed.

The approach of this angry army of miners caused consternation all round. Rumours flew unchecked through the streets, and some people advocated non-resistance. Meanwhile the anti-pikes were advancing. They tore away the Ashton pike, and proceeding to Bedminster, attacked the house of a justice of the peace; then, satisfied so far, they departed to Totterdown where the turnpikes were made to assume the appearance most suitable to the locality.

Thus far no attempt had been made to resist the rioters. They had gone as they pleased; the "police," or other guardians of the city, were alarmed, and declined to come into collision with the miners. But a valiant citizen, one Brickdale, stood forth like Horatius in the gap to keep the town, and assembling some other stout men and true, attacked the

rioters and took several prisoners. Two of these men were
tried and hanged at Taunton.

The others were sent up to Salisbury, where, says the historian,
"although the facts were notoriously proved against them, the
jury, being country people, would not find one of them guilty."

This did not end the rioting in the West. The Gloucestershire "pikes" were also demolished, and the perpetrators of
these outrages remained on the spot to demand money from
travellers! Could anything more clearly demonstrate the absurd
inconsistency of these patriots? They objected to turnpikes
because of the tolls levied by the trustees. The objectors first
destroyed the gates, and then demanded the tolls for themselves! Logicial truly!

Many conflicts occured, but the miners and sympathisers were
dispersed, and the pikes were replaced. But the evil had
spread. The Welsh in after years imbibed the same hearty
distaste for these artificial barriers. The tolls had become so
heavy that the country people determined to remove the cause
altogether, and in the autumn of 1839 an anti-turnpike league
was secretly formed and curiously managed. The magistrates
were making many improvements in the roads, and to do these
every road had its gate, so the tax upon travelling was extremely heavy in any case, but to anyone travelling into distant
counties the cost was very great.

Consequently a band of men united themselves, and more
in the spirit of frolic than in spite, in open day they demolished
some of the most obnoxious gates, which they considered had
been illegally erected.

The magistrates would not permit the trustees to replace
these gates ; and this concession to popular taste encouraged
the band. In 1843 the proceedings began in Caermarthen-
shire, and spread into the adjacent counties. The followers
of "Rebecca," the name adopted by the leader, worked in
secret but surely.

The origin of the title is an application of a text from Gene-
sis, which runs thus :—

" And they blessed Rebekah, and said unto her, Thou art our sister, be thou the mother of thousands of millions, and let thy seed possess the gate of those which hate them."

The result of this adaptation was most remarkable. Not a village, scarce a hamlet, in the counties named but had many adherents of Rebecca, who with her followers attired "herself " in female costume, and attended by many desperate characters on horseback, riding man-fashion, attacked and demolished every turnpike-gate.

The leader was a man of address and of some respectability. Certainly his arrangements were well made, and the marches made by night were excellently planned. Many wayfarers were surprised by meeting a troop of "horsewomen " on the lonely roads, and by hearing the sound of axes and hammers resounding through the still night air.

The farmers of the various districts were fully in sympathy with the Rebeccaites, and were " permitted " to provide the labour required to demolish the turnpikes. They were only pickling a rod for their own backs. " Rebeccas " proceeded as follows. So soon as they arrived at the pike, horns were blown, guns were fired, and the unfortunate toll-keeper was roused terrified from his sleep. The "females" began to cut away gates and posts and to pull down the toll-house, from which the affrighted keeper was compelled to hasten in whatever dress he could snatch up, and look on while his furniture was pulled out into the road. One good trait of the Rebeccas was the care they exercised in the removal of the furniture and person of the toll-keeper.

The gate destroyed, the riders would mount and gallop away to other quarry ; but when day came no trace of their whereabouts could be discovered. However, impunity begat some serious demands. The men found that they had "grievances," and insisted on their being redressed. They began to commit violence, and to attack "obnoxious " people, that is, those who did not agree with them. A clergyman has testified that he and his wife were in the habit of placing a wardrobe against their bedroom window every night "as a protection against fire-arms. His life was repeatedly in danger, and his curate was nearly murdered by a disguised party." Threatening letters were common, and black-mail a regular fine.

After a while the Welsh people found that the "whips" of the Government were mild in application in comparison with the "scorpions" of Rebecca. She was a true tyrant, selfish and oppressive, demanding money and free quarters, and ruining business. At length she went too far. Entering Caermarthen in force one day, "Rebecca" and her daughters attacked the workhouse and tried to pull it down. While thus engaged, pitching furniture, etc., into the street, the military, who had been hastily summoned, arrived. The mob and the dragoons came into collision, the former was signally defeated, and dispersed without any serious results, fortunately.

But, unfortunately, Rebecca would not take warning. She disgraced herself soon after by an attack on the Hendy gate, between Llanelly and Pontardulais, which was kept by an old woman whom the leaders had often warned to escape. She clung to her house, however, with the tenacity of age; and about three o'clock on Sunday morning, a number of the Rebecca gang set fire to her home.

The poor old dame rushed out, and bravely hastened to a neighbour, calling for assistance to subdue the fire. The person supplicated was afraid to move, and the old woman went back alone to save her property, when some one wantonly fired at her, and shot her through the chest. She staggered a short distance and fell dead!

This atrocity aroused the Government, and they took strong measures; but the feeling of the country may be guessed from the terms of the verdict of the coroner's jury, who brought in the finding that "The deceased died from effusion of blood into the chest, which occasioned suffocation; but from what cause, is to this jury unknown!"

The small disturbance in 1839, when in "fun" the turnpike was broken down, had become a national danger. No highway was safe, no homestead secure, no farmer unmulct. The Government alone could cope with these men who made Wales pandemonium ; who laughed at law, and couldn't abide order.

Soldiers were thickly picketed throughout the counties, and a strong body of the " A " division of Metropolitan police also arrived to keep order. But they did not immediately succeed. Rebecca came out in the dark and cut down the gates as before ; but intimidation ceased. In addition to this the trustees of many gates decided not to replace those broken down ; and finally a Commission was sent to investigate the grievances of the people.

These stringent and pacific measures did much to quiet the inhabitants, who gradually retired from the Rebecca associations. The Turnpike Laws were investigated, and tranquillity was restored. Poverty and the numerous imposts had been the united causes of as singular and as damaging an association as ever grew out of private, unpolitical grievances.

BRISTOL, FROM AN OLD PRINT.

COACH AND HORSE-SEDANS OF THE SIXTEENTH CENTURY.

CHAPTER VII.

TOLL-FARMERS.—ROMANCE OF THE GATE.—MAIL-COACH PRO-
CESSION.—POSTING AND POST-BAGS.—HACKNEY COACHES.

THE pike was scotched, not killed. It remained
for many a year after to bar the progress of road
communications. Farmers of tolls made large
fortunes, became hackney-coach proprietors and
" carriers," or the letters of coach-horses. The gates were held
under certain trusts, though they gave none ; and we can most
of us remember the long list of tolls which set forth the required payment, from the one penny of the pedestrian to the
shilling of the conveyance. I remember on one occasion my
facetious companion tendered to the toll-man threepence for
my passage, on the ground that I, his friend, was a donkey !

The man did not take the threepence; he was a discriminating person, *I* thought.

Many romances cluster round the old toll-bars; tales of those who have risen from assistants at the gate to be owners of many tolls and of untold wealth. One we recall—a true story—which is most interesting.

One day, a lad selling oranges by the road side, performed some trifling service at a certain gate in Essex for a wealthy banker. The gentleman, as he passed on, threw the lad a coin

RETURNING TO SCHOOL ON "BLACK MONDAY."

which proved to be a guinea. The lad, surprised, took it to the keeper of the gate, who declared that the gentleman, whom he well knew as a banker, must have made a mistake; and he advised the lad to remain and return the money to the donor in the evening.

The honest lad agreed, and waited all the day until his benefactor came back again. To him he explained the mistake, and the banker, much pleased, bade him keep the guinea, and also recommended him to the gate-keeper. The latter, willing to oblige a powerful man, took the lad into his hut and let him sleep there; subsequently giving him night work to do. The lad gave satisfaction, and received occasional " tips " from the banker.

After some years the old gate-keeper died, and the lad thought he might succeed him. He immediately applied to the benevolent banker, who recommended him to the trustees, and the lad was rewarded with the appointment in due course, greatly to his satisfaction.

Time passed, years rolled on : the young man had saved money, and had become rich and middle-aged. He had farmed well, and was reaping an abundant harvest from his hackney coaches and his gates. Business in the city was dull, and crashes at that period imminent. No one was safe ; the richest might be dragged down by no fault of his own, and credit was tottering.

One afternoon the eminent banker, who had resided in Essex, was seated in his parlour. To the outer office came a man unknown to banks and bankers, a somewhat shabby man the clerks thought. He wanted to see Mr. M—— the senior partner, but was put off with an excuse, he " wasn't in."

" Very well, then ; I'll wait till he comes," replied the visitor.

He was allowed to wait, which he did with the greatest patience until at length one went into the great man's room and informed the banker that a man had been waiting an immense time to see him, and still waited.

"Who is he ? "

" We do not know, sir ; he gave no name."

" Let him come in," said the banker ; " I will see him."

The visitor was admitted, and bidden to be seated.

" What can I do for you ? " asked the banker.

"You don't remember me ? " said the visitor. "Well, do you recollect giving a lad a guinea many years ago, at the I—— turnpike ? "

" Yes " ; after a pause the banker recollected the incident.

" And you recommended me for a post of gate-keeper,"

continued the man. " Now, these are hard times for bankers. You did me a kind turn, sir, several years ago. Now look here."

He put his hand into his breast pocket, and continued :—

" Here are thirty thousand pounds ; take them, put the money into your business. We can never tell what may happen. It may save your house these times——"

" But," exclaimed the astonished banker, " I cannot take your money, I may lose it ! "

" Well, then, I shall not be ruined. Take it; do as you please with it, it is yours."

After some parley the loan was accepted. No disaster fell on that bank then, and the " Mouse," who had opened the turnpike gate, helped the " Lion " to some purpose.

It is not often that we meet with such a romantic story in real life, but the foregoing is true. No names are mentioned because the relation of that lad still lives in the odour of good deeds and kindly offices. When the same lad was older and richer, he was riding in a hackney coach, and stopped it in St. Paul's Churchyard as he beheld a boy crying.

" What is the matter, my lad ? " asked the gentleman of the lady who was trying to pacify the child.

" He wants a ride in a hackney coach," she replied ; " I cannot give him such a treat."

" Well, let me see. Tell me who you are ? " He took the boy into a bun-shop and solaced him while he interviewed the lady. Then he drove off, instituted inquiries, and next day, being satisfied, sent a hackney coach to the lady's house for the promised drive. The boy proved a steady, excellent fellow, and when the old turnpike owner died he left him a large fortune, which he still enjoys ; and not only so, but makes happy all with whom he comes in contact. Long may he live, say all who know him as a genial, kindly gentleman.

Dear me ! These reminiscences of pikes come thickly. Who remembers the pea-shooters from the loaded coach topped and stuffed with schoolboys, which drove such a hail of peas on the toll-man's head and face ?   Who can recall the time when the coach was arrested at a gate because the keeper was in arrears with his money, and as the guard was forbidden to pay until the amount had been handed to the trustees, the keeper took his revenge, and shut the gate which was not opened until payment had been made ?

Who of us can remember the cutting down of a gate by the guard and passengers to circumvent a saucy pikeman ?   Who can remember Joseph Baxendale, so renowned in coaching annals, who founded " Pickford and Co "? and can any one recall Mr. Chaplin, who horsed the coaches after having driven them for some years ?"   Of " the twenty-seven mail coaches leaving London nightly he horsed fourteen," says Mr. Harris,* and they were those to " Holyhead, Devonport, Liverpool, Manchester, Bristol, Halifax, Portsmouth, Norwich, Hull, Bath, Dover, Poole and Southampton, Stroud, Lynn, and Wells."   He subsequently joined the London and Southampton Board, and established a partnership with Mr. Horne, of the Golden Cross.   His head-quarters was at " The Swan with two necks."

Edward Sherman and Louis Levy also figure as proprietors. The former, according to " The Coaching Age," had his quarters at the old sign of " The Boulogne Mouth," now corrupted from the harbour-bar to " The Bull and Mouth," a somewhat absurd change, as no bull ever existed without a mouth; hence the sign is in its adulterated form meaningless, if not explained by a man in the jaws of a bull.

Again, who can remember the great procession of the mail coaches on the king's birthday, or that which quitted the post-office nightly ?   The former assembled in Lincoln's Inn Fields, like a four-in-hand "meet," and passed through the principal streets as an exhibition.   This was an imposing sight with the scarlet-clothed guards and postmen.   The writer can remember the scarlet uniforms of the letter-carriers who looked so

* " The Coaching Age."

smart on the Queen's birthday, as smart almost as the epau-
letted guards (Grenadier) in their "claw-hammer" coats.   In
those old days the army had to make way for the mails, and
the coach drove through their ranks, as Mr. Harris anecdotally
reminds us.

In the entertaining "Annals of the Road" we meet with an
excellent description of the procession of mails on the King's
birthday.   The coaches were new each year, as at present to
the military and other services is issued new clothing on the
Queen's birthday.   The coaches filed out from Mr. Vidler's
yard in Millbank, filled with the friends and relatives of the
coachmen and guards, who wore immense "nosegays," beaver
hats with gold lace and cockades.   The key-bugles played
merry airs, and when a coach arrived late the tune was
changed to "O dear, what can the matter be!"

The bells rang when the procession started, the oldest
established mail coach in front.   The long line of vehicles,
headed by the inspectors of mail-coaches on horseback, filed
past St. James's Palace where the King and Queen, the Duke
of Richmond, Postmaster-General, and the Duke of Wellington
were standing to receive the respectful greetings of the coachmen and guards.

In former days, before 1829, the General Post-Office was in
Lombard Street.   The passengers and luggage of the mail-
coaches were strictly limited even in 1833, when only three
passengers were permitted outside.   The guards were well
paid and made as much as £4 to £5 a week, for though their
pay was absurdly small—some 10s. a week—the "tips"
amounted to considerable sums.   As before remarked, the
guard was literally a guard, armed, and to his care were many
valuable packages entrusted.   The mail-coaches were embel-
lished with the royal arms on the door and the "Orders" of
the Bath, St. Patrick, the Garter, and the Thistle on the
panels.   Sometimes sad accidents occurred; coaches in winter
were frequently plunged into heavy drifts, and extra horses
had to be ridden as leaders, to pull out the vehicles.   In the

coaching days of old the mail was the conveyer of news.
When a victory had been gained in the Peninsula, the coachman, guard, and coach were decorated with oak and laurel, the
horses with ribbons rare. Readers of Lever's books will remember that when the immortal Mickey Free came home
with his master, Charles O'Malley, he had the announcement
of the capture of Cuidad Rodrigo pasted on the post-chase
behind.

De Quincey describes the enthusiasm which was displayed
after Talavera, as the newspaper was unfolded, and the tale of

MAIL COACH.

the "glorious victory" told or read to the eager listeners.
Guards and coachmen in royal liveries, without overcoats—
"such a costume, and the elaborate arrangement of laurel in
their hats, dilate their hearts by giving to them openly a personal connection with the great news."

We have not space to detail the Romances of the Road in
connection with the mails, nor can we dwell on the humours of
the highway which are set forth in coaching annals and in books
published by writers of the time.. To Charles Dickens we are

indebted for many a glimpse of coach-life, and in "Sketches by Boz" we have the description of the starting of the coach graphically portrayed.

On May Day the coaches raced, not with each other, but "against time." This trial of speed compelled the absence of passengers and baggage; and in 1830 the independent "Tally Ho," which ran between London and Birmingham, performed an unparalleled feat, having travelled the distance of 109 miles in seven hours and thirty-nine minutes.* This was considered, as indeed it was, fast time before the railroad came into use.

The coaches under ordinary circumstances quitted the Post Office at eight o'clock, and about that hour some twenty-one well-horsed vehicles were despatched on their several ways, north and south, east and west; and a marked feature of the arrangements was the wonderful punctuality with which the coaches met and arrived. From Palmer's time until, perhaps, 1843, or thereabouts, some sixty years, the coaching age may be said to have lasted. Our highways were the witnesses of many amusing scenes, of many serious accidents, of many romantic occurrences.

The old inns too have gone the way of the coaches. The names of many still survive: "The Bull and Mouth," "The Swan with two necks" ("nicks"—the Vintner's mark), "The Belle Sauvage," "The Golden Cross," "The Saracen's Head," "The Bolt in Tun," etc. These are still familiar names as receiving offices for parcels and luggage, and form a link between the highway and the railway.

What anecdotes could be related of our roads in connection with the post-boy, that peculiar and ever youthful person who descended from the position of the mail carrier to that of postillion. The curious tenacity of life attributed to these gentry was commented upon by Mr. Samuel Weller, who compared the post-boy with the long-suffering donkey in his strong attachment to life.

* "Annals of the Road."

The post-boy has been frequently described; Cowper described him as the

> " Herald of a noisy world,
> With spattered boots, strapped waist and frozen locks,
> News from all nations lumbering at his back ;
> True to his charge, the close-packed load behind,
> Yet careless what he brings—his one concern
> Is to conduct it to the nearest inn,
> And having dropped th' expectant bag, pass on."

No doubt the fellows were most independent, but trustworthy. They did pretty much as they pleased, and were allowed to ride at a very moderate rate indeed. So far from carrying the mails with express speed or post haste these gentlemen could officially dally at the rate of five miles an hour. They accepted gratuities and "drinks," and in short did pretty much as they liked. The American Pony Express was the last relic of these "riding days."

The Pony Express was a wonderful institution. In the spring of 1859 the Pike's Peak Express Company established a stage line between the Missouri River and the Rocky Mountains. The distance—seven hundred miles—was made in six days and nights. Those days were full of adventure, as all the world, his wife and family were setting westward to join the gold-hunters. Wells, Fargo & Co. afterwards obtained the "stageing," but in 1860 the Pike's Peak Company arranged the celebrated Pony Express which established a rapid mail communication between the inhabitants of the eastern and western provinces before the telegraph had penetrated westward.

The mails used to be sent round by San Francisco from New York. The railway stopped in those days at the Missouri, and thence to the Pacific border did the ponies run. This was an enterprise which was, and is, considered marvellous. Indians, and road agents—as highwaymen were generously and euphemistically called—were continually on the plains, across which no road ran, and scarce a track indicated the route.

G

Bold, brave men and splendid cattle were demanded and obtained. Think of the conditions of the service ; imagine the ride through the vast solitudes of the far-reaching plains to the small log hut where, if the expressman arrived in safety, he found the relay mounted, or standing bridle on arm beside his wiry steed, ready to mount and ride across the next patch alone ; rushing into danger, mayhap to death ; or if not in actual bodily peril himself during his ride, he might reach the way station only to find it a heap of ashes and charred logs, the stable empty, and only his tired pony to rely on for the next " stage."

Under such conditions were the two thousand miles covered. No delays were allowed ; the bag or bags were never still. At every station the horses were changed, the riders at intervals of fifty or seventy miles.   The men rode gallantly, fearlessly, and many an exciting chase, many a siege, did the brave express men endure while urging on their wild career over the solitary plain, or while " resting " and awaiting the mails.

When the telegraph was put up in 1862 the Pony Express was discontinued ; but the speed of the service induced the Government to send the mails by land, by stage (and afterwards by railway), in preference to the steamer.   So the American post-boy died out and became the driver of the " stage " up and down the rough roads to Denver and through Colorado to the western ocean.

Mais, revenons a nos postillons.   The British post-boy rode quickly, but underwent many perils in his course.   In those happy days the highwayman was a professor, and robbery a profession.   On such roads as we have already described the post-boy had to ride with the mails, and was often robbed, sometimes maltreated.   The letters, etc., were not numerous, as we may judge, for one lad carried the mails for thirty-four towns in England, besides the Irish mail on one occasion on which he was robbed in the year 1779 " between Euston and Shipstone " in Oxfordshire.   What would the " Wild Irishman " mail train think of the bags of those days ?

From carrying the mails the post-boys came to riding post-horses with the post-chaise, which wealthy people hired. These post-chaises were supplied with relays at certain posting-houses—inns—at which the mails seldom stopped. The coach office and commercial house, or inn, went together as naturally as the hotel and posting-house, which can be seen unto this day. The distinction was perfectly understood near the metropolis, and at the posting houses the "boys" were always in readiness to ride the stages and drive the post-chaises home.

MAIL COACH STARTING FROM GENERAL POST OFFICE.

It appears from the statements in "Old Coaching Days" that the boy always drove the chaise home when it was empty. The riders had stable-assistants who were known as "cads." The average number of changes of horses was about twenty-five. The boys always went out in rotation. There was in the school-days of the present writer a very popular song for bedroom singing, termed the "Three Jolly Postboys," which had a fine chorus, and was as popular as the "Thoroughbred Oxford Man."

"Drink, boys, drink, boys, drink away sorrow,
For perhaps we may not drink again to-morrow '

was the chorus, with

"Come, landlord, fill the flowing bowl
Until it doth run over,
For to-night we'll merry be (three times),
To-morrow we'll be sober ! "

A great many years have elapsed since the dormitory re-
sounded with those strains, and memory may be at fault, but
the general "drift" of the song was "drinking at the Dragon,"
and many of the sentiments were repeated to eke out the
melody, which seems to me to have been a cheerful variation
of "Away with Melancholy." . . .

A ticket was handed to the post-boy when he set off on his
ride, and it was given up at the turnpike. The orthodox post-
boy is not frequently seen now-a-days, but we remember the
white beaver, the yellow or blue jackets, the cord tops and
"jockey" whips of the small, aged-looking "boys" who used
to ride postilion. Though posting was expensive in the past
it was greatly indulged in, and heavy payments were made to
the trustees of turnpikes. At one gate, says Mr. Harris, on
the Brighton road, the tolls amounted to £2,400 a year, of
which sum two-thirds appears to have been paid by coaches ;
but £800 was not a bad return for posting at one gate in one
year. Mr. Levy, we are informed, "farmed tolls to the amount
of £400,000 or £500,000 a year, and post-house duties to the
amount of £300,000 or £400,000 a year.* Turnpike bonds
were good investments in those days.

Before concluding this chapter we should say something
about the old hackney coaches, of which the derivation is so
little known. Many people imagine that those vehicles were
called so because they were originally made at Hackney, near
London. This is an error, and similar ignorance prevails

* " Old Coaching Days."

concerning the derivation of the French term "fiacre." We have investigated these cases, and have ascertained that the term "Hackney" is derived from the French.

Hackney coaches were first in use in England in 1605, before the stage-coach, which did not come into vogue until some thirty-five years afterwards. The name was derived from the French word *hacquenée*, signifying the horse that was hired to ride from stage to stage, or to draw carriages and private coaches. The word "hack" is the contraction of the old term. Readers of Pepys will find the word *hacquenée* in his writings.

Fynes Morryson wrote in 1617 that there were post-horses in England at stations about ten miles apart that could be hired by travellers on horseback at the charge of $2\frac{1}{2}d$. to $3d$. per mile, but "most travellers ride their own horses," yet "coaches are not to be hired anywhere but in London."

The hackney coach was introduced to the French by M. Nicholas Sauvage, who resided at the sign of St. Fiacre— hence the Parisian term for hackney carriage. The cabriolet was introduced on the 23rd April, 1823. This was a two-wheeled vehicle with a hood ; a lad or man used to stand on a board behind as "young Bailey" did for Mr. Montague Tigg. The first pair-wheeled cab was something of this kind, which a Mr. Davies introduced. These cabs were made to hold two persons besides the driver "who is partitioned off from his company." There are pictures of these cabs in many volumes, the driver seated at the side, a suggestion which was improved upon by Mr. Hansom, whose "patent safety" has been termed the "gondola of the London streets."

There can be no improvement upon the Hansom cab ; it is at present, with its easy springs and rubber tires, the most pleasant mode of conveyance we possess. Yet not long ago it was considered very indelicate, and certainly a loss of caste, if a lady rode in a hansom alone.

Of the useful omnibus Mr. Shillibear, a coachmaker, was the introducer. The first fares were one shilling from Lisson

Grove to the Bank.   These omnibuses were large vehicles, carrying twenty-two passengers inside, no "outsides" then, and no knife-board.   This seat was added in 1849.   Readers who may be interested in the "History of Coaches" will find plenty of information in Mr. Thrupp's work, to which we are indebted for some of the foregoing information regarding cabs.

By degrees coach and post travelling declined; but the former has already reasserted itself, the latter mode is chiefly confined to the advocates of "driving tours," who do not wish to take out their own horses.   But the Romance of the Highway may be now declared closed.

AN OMNIBUS OF THE PRESENT DAY.

A VIEW IN THE FEN COUNTRY.

THE ROMANCE OF THE WATER-WAY.

CHAPTER I.

CANALS AND THEIR ORIGIN.—THE FENS AND THEIR STORY.

EVERYONE who has travelled in the English Midlands, in the Netherlands, or France, is acquainted with the net-work of canals which in places vies with the more modern railway. One can even travel not uncomfortably through the northern districts of London by canal, and reach the Thames near Limehouse.

From a very early date, more than six hundred years before the Christian era, these water-ways were in use in China and in Egypt. It is a curious fact that of nearly all nations England, lately so much in the forefront of all great engineering enterprises, should have been the last to embark upon the construction of canals. This seems, perhaps, even more extraordinary when we consider that other means of communication were in a very rough and unready condition. We have already pointed out the state of our roads in the eighteenth century, and in consequence of the difficulty and the expense of transit, even by pack animals, produce was not carried from place to place by land : the necessary supplies were brought from abroad, or coastwise, in ships.

"Domestic commerce was simply impossible," says Mr. Smiles, and he then goes on to enumerate the articles brought from abroad, for "it cost much less to bring goods from Hamburg, Amsterdam, or Havre, by sea, than from Norwich or Birmingham by land."

This condition of affairs, not much more than a hundred years ago, may be considered impossible ; but the history of England, the social history of that time, and the subsequent records of the present century prove beyond a doubt that all, or nearly all, our great enterprises have been initiated and undertaken within the present century, and if we put the age of our progress at one hundred and twenty years, we shall be about correct, for it was in 1776 that James Watt made the steam-engine a working machine. Since then mechanical science has advanced by leaps and bounds ; but it must be admitted that both Smeaton and Brindley had been at work before the date quoted above.

Yet it was not for want of example that scientific men did not arise. There was plenty of talent in England, plenty of foreign talent. But it must be confessed that the upper class Englishman 'n the middle of the eighteenth century looked down upon such work as "engineers" performed. It is well known that even Smeaton was derided for road-making, and

when wars and disturbance took away the foreigner, or ban-
ished him from our island, "some one" had to do these
"jobs"—and the most practical working-men did them.

France likewise boasted her engineers, but one of the
greatest of them was a self-taught man who made a grand canal
nearly a hundred years before we English had a canal at all!

JAMES WATT.

Italy, France, Holland especially, and Russia, had paid attention to canal navigation long before England, and the first
great impetus to inland communication came from a disappointment in love!

Nor could ignorance have been pleaded by our countrymen
as an excuse for not entering upon such useful works. We
had the examples of the Dutch, Flemish, and French engineers

in our midst ; and of Peter the Great and Italian workers long

AQUEDUCT AND ROAD BRIDGE AT NIMES.

before. Our Sir Hugh Myddleton himself had completed a
splendid work which, though not a canal, was an example of

what men could accomplish in the way of water-works, and John Perry, Peter the Great's engineer, came to the rescue and filled the breach in the Thames at Dagenham.

Besides these examples we find, still farther back, certain efforts made to open up inland river communication. One John Trew, as unlucky as John Perry, made a small cut from Exeter to Topsham, in the time of Henry the Eighth (1566). About one hundred years later we find another attempt was made to open up the rivers Avon and Isis; and a work on the "improvement of England," published in 1677, gives details of many schemes. The first private water supply in this huge London of ours was initiated by Morice, a Dutchman, in 1582, and since then the question of the Metropolitan water supply has been continually before the British public, even as it is at the present time of writing.

Nor is it surprising that the question of water supply should be considered so important. There is no element of our civilization so important as water, which is the first necessary of life. Nature and humanity, the whole world, would be dead without this liquid, even as the moon at present : a wreck of a planet without atmosphere, vegetation, or life in any form in keeping with our ideas.

These are truisms, commonplaces which we all acknowledge, and as usually forget ! We do not consider that water composes by far the major portion of our mortal frames, and that it is being always " pumped up " by the sun, evaporated, purified, and sent down again to the earth, to go through the same round of rain, snow, ice, glacier, and spring water, to cut away rocks, change the features of our globe, and sustain the inhabitants thereof, so long as the world shall last.

The Romans—those knowing ancient Romans—knew the value of this best of Nature's gifts, and built long-lasting aqueducts. We are all familiar with these monuments of Roman skill. That same Appius Claudius who gave his name to the Appian Way, also constructed the first aqueduct in Rome and called it Appia Aqua. Even as the victorious

Romans made their roads, they left behind them aqueducts Look at those graceful structures, with numerous arches, perched so lightly upon other arches, sustained in mid-air between the lofty banks, upon which is carried the water supply first bestowed by the Romans.   Go to Nimes, or Segovia, and see what engineering was in those years before the Christian era.   Julius Frontius declared that the aqueducts were the most distinguishing marks of the Roman emperors, his masters.

This section of our work might be divided into sub-sections, viz., Inland Navigation, Water Works, Water-Distribution and Supply, and lastly Docks and Locks, if space permitted.   The use and importance of water as a motive power, even in these days of steam and electricity, is more and more in demand, and the parent of steam even displaces its more celebrated offspring at times.

There are few incidents in history more familiar to us than that in connection with the worthless King John, when he lost his baggage in the fen country.   We are all familiar with the Isle of Ely and other so-called islands in the east of England, but we do not, as a rule, consider why these places have had such titles bestowed on them.   To the youthful reader the name carries no historical meaning, and yet the Lincolnshire fens are connected with the "very first beginnings" of inland navigation in England.

Eye, in Suffolk, was also anciently an island, and Leland states that "in old time barges came up thither from the haven of Chromer, or some creek near it"; and old Sir William Dugdale gives us a very interesting account of the reclamation of these overwhelmed districts.   His book, printed by Alice Warren in 1622, contains the best information to be found on this early engineering, and we will refer to him for it.

The Romans, it appears, were the first who made any attempt to reclaim the districts of Holland, and Marshland (in England), near the Wash.   The land thus kept from the sea became "a fenny lake," in consequence of the rivers which flowed into the immense low-lying tract. To drain this the Romans constructed dykes, and the Caer Dyke is the oldest "cut" in England.

Much romantic and historical interest hangs about these isolated tracts of land amid the lakes and fens. Wittlesea Mere was then a lake, and we read that Danish transports sailed over the fen country. The islands of Ely, Thorney, Spinney, Crowland, Ramesey, and others were inhabited by holy hermits to whose abodes "there is no access but by navigable vessels, except unto Ramesey by a causey." This causeway was raised above the mere, and was doubtless of Roman origin.

The horrors of this "hideous fen of a huge bigness" are given in the picturesque old style. "Oft times clouded with mist and dark vapours, having within it islands and woods, as also crooked and winding rivers." Thither retired Guthlac, the founder of the Abbey of Croyland, and the builder of the quaint bridge which we have already mentioned in the foregoing section of this work. The bridge spanned three streams, now, of course, dry. In Guthlac's solitary days he was accustomed to discern his cell to be full of black troops of unclean spirits, which crept in "under the dore."

What occasioned these "black devils" the chronicler does not state. They were very terrible in appearance, with horse's teeth, and distorted, swollen limbs, of hideous mien, and awful in the way of features. These fen-fiends used to drag the holy man from his bed, and cast him into the dirty water, tearing his limbs by drawing him amongst brambles and briars! This is a terrible narrative, and leaves a loophole for the scoffer: but we do not intend to follow Guthlac farther. Peace to his ashes.

Etheldreda, wife of the prince of the fen-men, retired to the island of Ely, or Eel-island, after her second marriage to Egfrid, king of Northumberland, and as at Croyland an abbey was founded there. The same principle was adopted at Thorney and Ramesey. Croyland and those other fen settlements became famous; monks and nuns flocked thither, and

devotees came from all quarters.   Boats were employed, and
inland navigation began in the fens in "cuts" and "dykes."
In after times, when the drainage became better, the resident
monks turned road-makers, and the "abbot's delve," or ditch,
was made by the worthy Brithnod.   But Ely, like the other
abbeys, remained isolated and difficult of access for many
years first.

The isle of Ely became a refuge for fugitives.   Thus when

ISLAND OF ELY.

William the Norman conquered England, Stigand fled to the
Fens, and many others were defended by Hereward, whose
exploits are so romantic, but not in place here.   By successive
monarchs Fenland was taken in hand; commissioners were
sent down to drain and improve it.   James I. appointed Lord
Chief Justice Popham, Sir Thomas Fleming, Kt., Sir William
Rumney, alderman, and John Eldred, citizen, to drain all the
fens within the space of seven years at their own proper costs
and charges, and convey also the Ouse, the Nene, the Weland,

to make new banks and other improvements. The "undertakers" were to be supplied with such sums of money as "the Commissioners may think fit" from the persons benefited by such draining, and also as a recompense a large grant of the reclaimed land, with certain rights of fishing.

But notwithstanding all attempts the water had for a while the best of it. Embanking and dam-making are serious operations even now, and in those ante-puddling days most heart-breaking. The work went on very slowly. The seven years of the contract were extended to ten, and so little advance had been made in those early engineering works, that people scoffed and made rhymes and libellous songs to disparage the work, which had already been hindered by some perverse persons, just as the railroad, centuries after, was opposed by landowners.

One of these songs ran thus in 1611. It commences :—

" Come, brethren of the water, and let us all assemble
To treat upon this matter which makes us quake and tremble,
For we shall rue it if't be true that Fenns be undertaken,
And where we feed in fen and reed there'll feed both Beef and Bacon."

Then, after bewailing the design, the bard proceeds :—

"Away with Boates and Rodder, Farewell both Bootes and Skatches,
No need of one nor t'other, men now make better matches,
Stilt makers all and tanners shall complain of the disaster,
For they will make each muddy lake, for Essex calves a pasture.

The feathered Fowles have wings to fly to other nations,
But we have no such things to help our transportations.
We must give place, oh, grievous case, to horned Beasts and Cattell,
Except that we can all agree to drive them out by battell.

Wherefore let us entreat our sweetest water Nurses
To show their power so great as t'help to drain their purses,
And send us good old Captain Flood to lead us out to battell,
Then two-penny Jack with skates on his back will drive out all the
Cattell."

After eulogising good Captain Flood the ballad invokes the heathen deities, thus :—

"God Eölus, we do thee pray that thou wilt not be wanting,
Thou never said us nay, now listen to our canting,
Do thou deride their hope and pride that purpose our confusion,
And send a blast that they in haste may work no good conclusion.

Great Neptune (God of seas) this work must needs provoke thee,
They mean thee to disease, and with Fen water choake thee.
But with thy Mace do thou deface and quite confound this matter,
And send thy sands to make dry lands when they shall want fresh water.

And eke we pray thee Moon that thou wilt be propitious,
To see that nought be done to prosper the malicious.
The Summer's heat has wrought a feat whereby themselves they flatter,
Yet be so good as send a storm less Essex Calves want water."

It is strange that the inhabitants should have objected to
have their land drained and reclaimed from the ocean and the
river. Had they entirely forgotten the memorable tempest of
1571, when the "country of the stilt-walkers" was swept, and
scores of vessels lost, bridges carried away by swollen rivers,
and one ship carried inland on the tide was wrecked on a
house, on whose roof the crew clambered and were saved?
Had the "stilt-walkers" forgotten the loss of sheep and cattle
and human lives which had so frequently occurred in the
watery Fens?

Apparently they had; they cared not, and the work progressed but slowly; what was effected was not done without
assistance from the Dutch. The Commissioners contracted
with Cornelius Vermuyden, Esq., "of the City of London," in
the year 1626, to drain the Great Level. This engineer had
been in England a few years, and had successfully accomplished
other works, damming the Thames at Dagenham—always a
weak spot—and draining Windsor Park. The Fens were still
undrained when Vermuyden was instructed by James I. to
proceed with the clearing of Hatfield Chase, on the confines
of Yorkshire. Charles confirmed the order, and the reclamation of an immense tract containing about 70,000 acres was
undertaken.

Within this tract, which lies between Yorkshire and Lincolnshire, is included the celebrated "Island of Axholme," formerly one continued fen, occasioned by the silt thrown up by the Trent with the tides of the Humber.[1] We learn from our authority that this silting obstructed the free passage of the Dun and the Idle, and forced their waters back, so that the adjacent higher parts formed an island—hence the name. This became a place of refuge and a stronghold, formerly a hunting ground, but in the time of James merely a flat expanse of water.

The Dutch engineer undertook the task at his own expense on condition of receiving one-third of the land recovered. He engaged foreign hands, and for a while success attended his efforts in every way. The feat was performed within five years, at a cost of £55,825, the waters being conveyed into the river of Trent through Snow Sewer and Althorpe River by means of a sluice which prevented the reflux at high tide. For these services Vermuyden was knighted.

This draining improved the district immensely. Whereas the country thereabouts had formerly been full of wandering beggars, but very few came afterwards. The people were employed in the weeding of the crops and other agricultural work which paid them.

The workmen from Holland settled down, and gave quite a Flemish character to the district. They built their church, and were joined by many Protestant fugitives, Walloons and French, and continued in peaceable possession until the beginning of the trouble 'twixt King and Parliament. In 1642 the natives uprose, attacked the settlers, demolished their houses, and broke down their fences.

Not content with this, these turbulent men and women pulled up the sluices, and, by letting in the tide, again inundated a portion of Hatfield Chase. The inhabitants were forced to swim away "like ducks," and men with muskets were stationed at intervals to prevent any attempts at lowering the gates. Even this did not satisfy the natives of Axholme,

---

[1] "Beauties of England and Wales."

for they pulled down the embankments, filled up ditches, permitted the "cattell" to get into the corn, wilfully emulating the carelessness of "Little Boy Blue," and generally "made hay" of Vermuyden's colony.

These acts, partly due to hostility of race, and partly to political bias, greatly damaged the reclaimed lands. The Parliamentary party occupied the territory, and behaved in their usual iconoclastic manner; occupied the "Dutch" town, and placed their horses in the church. Colonel Lilburne, in the true spirit of his politics, having ousted the true owners, seized some of the land for himself, despite his philanthropic declaration in favour of the rights of the commoners. But he was eventually compelled to relinquish his ill-gotten grounds—some 2,000 acres of common land in Epworth.

Let us now look on another picture. Sir Cornelius, who had long before parted with his interest in the reclaimed land, some of which had been re-submerged by the riotous natives, had turned his attention to the Cambridge Fens, or "Levels." On the 16th January, 1629, a "Session of Sewers" was held at Huntingdon, and a tax of six shillings an acre was "laid upon all the said Marsh"—the Great Level to wit.

To accomplish the draining Sir Cornelius Vermuyden was again in request, but the people would have none of him. They denounced him as an alien, and made application to Francis, Earl of Bedford, who had a large stake in the Fen country through the grant of Thorney Island made to his ancestors.

The earl responded patriotically. He made an agreement to drain and reclaim the land on consideration of receiving ninety-five thousand acres of it. With him were associated thirteen gentlemen adventurers "of high rank and respectability." In 1634 a charter of incorporation was granted them, in three years and a half the work was completed at an expense of upwards of £100,000, and the grant of land was allotted to the earl (1637).

Nevertheless, the work was not complete, and another

SCENE IN THE FENS, CAMBRIDGESHIRE.

Session of Sewers was held in 1638, when the earl's work was adjudged defective. Then the king himself determined to undertake the business, on condition of receiving an addition of 57,000 acres to those to be taken from the Earl of Bedford, which he had forfeited. The king called in Sir Cornelius Vermuyden, and asked his advice, but not much came of this, for about the same time Charles and his parliament fell out, and the feelings of the Fen people were against the king and his projects. Rumours were set afloat that all Charles wanted was money. Cromwell took up the cudgels for the Fen-folk, and as "Lord of the Fens" started on his career of popularity which almost carried him to the throne of England.

The effects of the dispute and civil war in the Yorkshire side of the Fens have already been mentioned. The Cambridge plans were deemed hopeless. Popular ignorant opinion had pronounced against the project, and nothing further was attempted until 1649, when the parliament restored William, Earl of Bedford, to the position his father had occupied. The boundaries of the Levels were settled, and the work progressed under the fostering care of the very man who had denounced King Charles for attempting to perform it !

On the 25th March, 1653, the Commissioners pronounced the Level fully drained, and the earl with his associates were awarded the 95,000 acres of the reclaimed land which had been promised to Francis, his father. The new "gentlemen adventurers" were nearly ruined, for the cost of the work was £400,000.

The "great Bedford Level" was afterwards, in 1697, divided into three. This second undertaking composed the formation of the new Bedford River from Erith to Salter's Lode; "Smith's Leam" was cut, and thus communication opened between Wisbech and Peterborough. Many drains and dams were cut and formed, so that navigation and communication were improved and assisted. The completion of all these important water-ways was celebrated by the Commissioners at Ely, where the gentlemen adventurers and all

others concerned attended service in the cathedral; and
Vermuyden took a prominent part in the procession, for to
him the earl had turned as engineer.   The Commissioners
had previously sailed down the new cuts and rivers, personally
conducted by Vermuyden, who made a little speech on the
occasion.

WHITTLESEA MERE AS IT IS.

The iron post marks the subsidence of the soil (8 feet 2 inches) since drainage.

But honour and glory were all that the Dutch engineer
reaped.   It seems that he had to sell all he had to pay his
men; and thus, while he was draining the Fens, they were
retaliating and draining him !   To such a low level did he sink
that he was compelled to appeal to Parliament for assistance ;
but whether or not he received it is an open question.   One
fact is certain, he quitted England in poverty, and died in

Holland, no doubt a disappointed and broken man. The engineer was not regarded then as now.

But if Vermuyden perished, the result of his labours remained. The Fens were to a great extent drained; and meadow succeeded water. Great efforts were made in the direction of further improvements, but they were not entirely satisfactory until Rennie, the English engineer, was appointed to look into the question in 1789, and later, when he began to drain the East and West and the Wildmere Fens.

The condition of the country will be easily imagined. Generally under water, or so swampy that no one could depend upon its stability, the district was haunted by miasma and fever. The inhabitants who still stuck to the soil— "clung" is not the word—were ill, their sheep had the rot, and their "farms" were covered with weeds and thistles. Huts of reeds and houses of rushes dotted the watery waste, and "Eden in reality" must have been the counterpart of the Fen country, in which Mark Tapley would have been "jolly" with considerable credit to himself.

Sir Joseph Banks initiated the scheme in which Mr. Rennie was employed, and "he lived," writes Mr. Smiles, "at Revesby, where he kept open house." In those days lavish hospitality was a virtue, and every one enjoyed Sir Joseph's. On one occasion when Rennie went to Revesby, he cautioned the butler not to permit his post-boy to become intoxicated during the evening, or he would not be fit to drive next morning. The custom of the time is well exemplified in the man's reply, "I am sorry that I cannot oblige you, sir. The man left here sober last time, but only on condition that he would get drunk next time. Therefore, sir, for the honour of the house I must keep my word; but I will take care that you are not delayed for want of horses and a post-boy."

The butler was faithful: the post-boy got drunk in comfort, and Mr. Rennie was carefully despatched on his way next day in good time.[1] In 1799 Mr. Rennie was requested to investi-

[1] "Lives of the Engineers."

gate and report ; and he proceeded to drain the land by means of low-water sluices and cuts, which permitted the water to run off at ebb tide, but effectually prevented ingress at the flood.

It is scarcely necessary to enter into a detailed description of Mr. Rennie's plans. He made the outfalls as low as possible, and thus got rid of the maximum amount of water from the land. "Catchwater" drains to secure the high-level rains also were a part of his scheme, which was very comprehensive and equally simple. It involved the cutting of many drains, and ended in reclaiming 40,000 acres of land. Readers who require the fuller details of these useful works will find them in the "Lives" of the engineer himself, and other works, reports and local chronicles. It is sufficient to state that Rennie fully succeeded, and when steam-pumps came into use the drainage became merely a question of time.

During all these draining projects and cuttings many interesting relics were brought to light, and the remains of ancient forests exhumed. Firs and oaks—the latter in multitudes, and of extraordinary size—were found about three feet below the soil. The roots were still in the ground, the trunks having been burned, as is evident ; the "ends of them being *coaled*," says the old chronicler. These indications point to the occupation of the Romans, who burnt the trees or felled them to deprive their foes of shelter.

Besides these relics skeletons of fish (sea fish) have been discovered ; and the many places indicating fish in their titles prove that the waters covered the districts. These fish bones have been turned up many feet below the surface of the ground ; and a second bottom has been discovered, eight feet beneath the supposed bottom of the lakes, of a stony character, with boats embedded in it by silt. More curious still was the discovery at Wittlesea of a meadow beneath the surface, with the grass lying in swathes, "as if first mowed "!

These "finds" were afterwards supplemented by the discovery of a smith's forge, silted in, with hammers, pincers, and even horse-shoes, all sixteen feet below the surface ! Curious old shoes *temp*. Richard II., horns of animals ; tan-vats. and other indications of the ancient occupation of this tract were found. This formerly dry and cultivated land was subsequently reduced to a marshy and fenny district in consequence of the silting up of the embouchures of streams, which overflowed their low banks.

In Dugdale's interesting old work will be found an account of other reclaiming acts in Kent and Middlesex, Suffolk and other counties. Romney Marsh he describes at length, and tells us much of its ancient condition and of its regulations. We are confirmed in the view that these low-lying lands—and mayhap all land—originated naturally from sea deposits ; then by the accumulations carried down by rivers, and so on ; the water giving and taking in places, robbing Peter to pay Paul, washing away a continent to replenish an island, or perhaps another continent.

Thus Holland is the outcome of the Rhine, which laid it down under water. The marshes near London, Plumstead, and Erith, and the Essex flats generally indicate the same silting-up process as at Romney and round there. Embankments were constructed along the river Thames, and Lambeth Marsh remains to tell the story of the district once at the mercy of the river and its tides. Moorfields deserved the name, as they contained " a great fen on the north side of the city, on which when frozen the young people go to play upon the ice. Some, taking a little run, do set their feet a good distance and glide a great way." So much for the sliding of the period !

This fen stretched, according to Stow, between Bishopsgate and the postern called Cripplegate to Finsbury and the Holy Well. It long continued a waste, but after Edward II. came to the throne a " causey " was made and a gate erected, so that the citizens might go " to Hildon and Norton." Moreover, the ditches of the city from Shoreditch to Deepe Ditch, by Bethlem to the Moor-ditch, were drained, " by means of which the said fen or moor was greatly drained and dried."

Subsequently sluices were made, and the water from the fen was carried into the " Walbrooke," and so into the Thames. Our author then speaks of the Lea marshes—the river of Lye, anciently called Luye,—the marshes of Wolwyche in Kent, of West Hamme and Elthamme; and tells us how people were

THE RIVER LEA AT TOTTENHAM.

compelled to put up embankments against the overflowings of the Thames and Lea. The Thames was a frequent offender; and, as we have remarked, Vermuyden was called in to repair the Dagenham breach, which he succeeded in closing for a while. But the most effective work of this kind in our river was performed by John Perry, the Czar Peter's engineer, who, coming home disgusted with his imperial master, was after a severe struggle successful in his encounter with Father Thames.

The story of his battle, and its ending, will make an interesting chapter, and we propose to deal with it in our next.

TOWER OF LONDON FROM THE THAMES.

THAMES POLICE EMBARKING.

CHAPTER II.

OLD "FATHER" THAMES.—ITS ROMANCE.—JOHN PERRY AND
DAGENHAM REACH.—THE THAMES VANQUISHED.

O write anything concerning the Thames in these days seems unnecessary, when so many thousands of trippers annually "run" up or down the royal river, once the most frequented thoroughfare of London. Those who every year seek the Thames in its upper or lower reaches trouble themselves very little indeed with its engineering, and almost as little with its romances. Few regard the mighty embanked course of the river as anything exceptional. Few people think of the mighty works undertaken and so successfully carried out in the eighteenth and nineteenth centuries, and those still mightier tasks completed by the Romans and the Britons in the course of the centuries long past.

Curious as it may appear to the ordinary traveller on the
Thames, our fine river may almost be designated a canal, since
as far as its navigation is concerned it is, practically, an artificial work.

In "Dugdale" we read a great deal of the history of the
imbanking of the Thames, that placid stream whose open
stretches, waving corn-fields, snug smooth backwaters, and
tumbling weirs in the upper reaches tempt so many to romance and sentiment in the pair-oar or the sculling gig! The
varying reaches, the trim lawns, the white-clothed residents,
the clear still water mirroring the weeping ash, the tall elms,
or even the half-sleeping swans, the dormer windows of the
houses on the banks, and the quaint old gables above them,
are all familiar.

Down the reaches we float lazily till we reach the business
end of the Thames—the steamboats, the barges and the
wherries. Here romance may cease, for the policeman in
his boat-cloak is afloat on the look-out for "pirates" or other
floating prey, ghastly perhaps, or only strange.

In olden time a few men were told off to guard the banks,
and then the many-handed fraternity, known by such queer
names as "heavy or light horsemen," "game lightermen,"
"jumpers," "scuffle-hunters," and such titles—all pirates and
thieves. "Loafers" flourished. Every one who had nothing
to do but to pick up what he could find tried the river, unchecked, even abetted by the officials of the Customs, who derived no small profits from the simple, natural acts of "wink-
ing," "shutting their eyes," and "holding out their hands."

So in those good old times these men, in parties organized
for the purpose, would board vessels openly, and steal anything,
—sugar (highly valued), tea and other consignments,—until it
was calculated that at least £70,000 a year was lost by these
malpractices. But the nineteenth century, the age of progress,
put a stop to this by a proper system of patrolling.

If one wishes to see the true present-day romance of the
water-way, he has to go out on the Thames in the police-boat,

if he can gain permission; or, if not, in a barge, which the writer can state from personal experience is by no means an unpleasant method of travelling.

The Thames police arrange their work in semi-nautical fashion by "watches," six hours in the twenty-four, and perhaps a dozen boats are patrolling. In this business they are compelled to peep into all kinds of creeks and crannies, much to the disgust of the watermen; some of whom *may* have the tastes of Mr. Riderhood, and may be "birds of prey" of the class depicted by "Our Mutual Friend," if any such ever existed. The police, though, can tell you of many sad cases of suicide, and of many rescues during the small hours.

It is rather wild work pulling up or down stream without showing any lights, and the risk of being "smashed up" is great. But the police-boat is handled with dexterity and intelligence, and the coxswain manages to slip in between the barges and steamers, and to peep into any suspicious craft, in safety. A single dark lantern beneath the thwarts suffices to aid the keen practised vision of the inspector, who, cloaked in the stern of the galley, threads his way up and down the Thames with unerring tact and certainty. The work is at times supremely trying. In fine weather it is not a bad occupation to pull up and down stream, but when the bitter easterly wind, the snow, hail and sleet are falling thickly, or a dense fog shrouds the great looming hulls of mighty steamers, which, blowing "syrens," come drifting along the dark waters, then the policeman's lot is emphatically *not* a happy one.

And so in the barge traffic, which proceeds up and down with business-like laziness on the flowing and ebbing tides. Sometimes a puffing, fussy steam-tug renders assistance; at times, also, sails are set, and then the barge is not dependent on her sweeps. In a barge one can inspect the formation of the river, once, if one may judge from the extent of the former marshes on its course, an extensive, meandering stream, sorely dammed by old London Bridge, and permitted to roam out of its course in many places in Surrey, Middlesex, Essex, and Kent. The first mention of this serious reclamation of these old marshes seems to have been in "the Eighth Ed. 2," when "John Abel and John de Hortone were constituted Commissioners for to vew and take order for the repair of the banks, ditches, etc. between Dartford Flete and Grenewich."

Another Commission was subsequently appointed to remedy the broken banks "betwixt Grenewich and London Bridge," to repair the newly-made breaches between "Grenewich and

GREENWICH FROM THE RIVER.

Wolwiche"; and that towards Bermondsey. Also to "oversee the banks betwixt Grenewiche and Plumstede, and betwixt Wolwiche and Earith."

Again we read of repairs, or new constructions to be built between "Wollwyche and Southwerke," in the time of Richard II., "according to Marish law"; and with authority to "imprest labor." By these acts and others, which will be quoted presently, we may fairly perceive that the embankment of the Thames was no new thing in those days; that "good husbandry" was far more ancient, for Southwark is built upon reclaimed or artificial soil, and was occupied by the Romans, as the ruins and remains of Roman pavements have plainly attested.

In the time of Edward I., William de Carleton had commission to repair the banks "betwixt Grenewich and Lambehethe," and by another Commission in the time of Edward II. repairs were ordained between London Bridge "and the Mannour called Fauxes Hall"; and for these purposes the cost was claimed from the owner of the property benefited. The sewers were also cleansed; one called Dyflete, from the water of St. Thomas into the Thames, "having been stopped up, and the water drowned the adjacent marshes."

Later, in Henry VI.'s time, we find an ordinance to view the banks, "within the lordships of South Lambehethe, North Lambehethe, Lambehethe Marsh, and Parysh-garden as in Southwerk, Bermondsey, Rotherhithe, Depfordstroude, Peckham, Hacham, Camerwell, Stokwell, Clapham, and Newyngton," then broken and decayed. And by another, "31 Henry VI.," some commissioners were nominated to view all the banks "from East Grenewiche in Kent to Wandelworth in Surrey." In these presents were also included "Batersey" and "Camberwerwell."

The town of "Stebenhithe" (Stepney) seems to have attracted attention, and the Bishop took an "Inquisition" at the Hospital of St. Katherine's, near the Tower of London, for ·iew of the banks and ditches lying betwixt the said Hospital and the town of "Chadewelle," or "Shadwell;" Berkynflete, in Essex, was (later) included, with those by Blakewall, Stratford-atte-Bowe, the "town of Lymeostes, and the Wall called Black Wall."

As already remarked, Moorfields were marshy, and Fleet Street neighbourhood swampy. Thus we have ample evidence to prove that the present Thames River is in its lower channel artificial, and that it must have been embanked by the Romans or other early residents in Britain. It is at any rate sufficient for our purpose to show the existence of those embankments, which were so often giving way, and permitting the river to drown the adjacent low-lying lands. As the districts and towns about the City became incorporated with it, the inhabitants "fortified" themselves against the oft-recurring attacks by opening the water-way—that is, by enlarging the bridges. Sometimes the channel was choked up, and then the reigning king sent a "presentment," and distrained on the owners.

In the reign of Elizabeth, Commissioners were appointed for the recovery of Havering Marsh, "then overflown and drowned"; and for preventing the like to Dagenham Level, it was decreed that Dagenham Creek should be immediately "issued"; and as the damage "had been caused by a breach in the wall of Will: Aloff, of Hornchurch, Esquire, he, the said William, to pay the sum of five hundred pounds, and the land-holders of Dagenham certain rates by the acre."

But this did not serve. In 1621 Cornelius Vermuyden was appointed to stop the breach, and succeeded in perfecting the work, for which, as the owners declined to pay, Vermuyden was awarded "a parcel of the lands" by the king's letters patent.

If we pleased, we could relate many other efforts to retain the land in other parts of England, but those quoted will suffice. The Lea, it may be remarked, was just as intrusive as the Thames. Limehouse was swamped in 1676, and the Isle of Dogs—the locality of the royal kennels—was submerged. In 1707 a tremendous breach occurred at Dagenham, and instead of being immediately closed, it was neglected. At every tide it increased, the water ran at its own sweet will, and finding a lower level where it could rest in peace, undisturbed by tides and ships, it settled down; and when the tides ran out, it permitted the soil to escape also. After a while the soil began to assert itself in the river in revenge. Turned out of the flats it rose into a bank, and held up its crest in the middle of the Thames.

When the people saw this, and heard of the danger to

commerce, they arose and tried to dam the breach in the embankment. But the water and tide derided all their puny efforts, their fascines and "chalk trunks," their piles and sunken vessels. The Government met with no success either. Seven years passed away, and the breach was not healed. Contractors came and went, but the tide flowed on, came up smiling as it were, and in a few hours crushed the caissons and beams, and sent them to the bottom of the river, or floating, as the case might be.

No one could tackle this business. Father Thames would not be any more coerced, and it seemed as if the river would remain absolute master of the situation, when Captain Perry appeared in London. The origin of the mischief arose from the bursting of a sluice for the draining of the levels, and could easily have been stopped at first.

It was about a fortnight after Captain Perry's return from Russia in 1713, that a gentleman connected with the attempts to stop the breach requested the engineer to go and see it. He found that the task was difficult, heard all the accounts of the mishaps that had occurred. Experts could stop the water, they said, but could not stop the ground; whereas, if they stopped the ground, Perry considered that they would keep out the water. The gallant captain, who had served his country afloat, seems to have been a cautious man. He would not give the workers any hints, but gave them the idea that the breach *could* be stopped! How, he did not say, and then he left them to examine the extent of the inundation.

During the fourteen years which had elapsed since the first opening in the bank was made, the river had extended into several branches "like arms of a river," the longest of which was above a mile and a half, and in places some 400 to 500 feet in breadth, and in depth from 20 to 40 feet. About 120 acres of land had been washed away into the Thames. When Captain Perry had made his inspection, and determined in his own mind not to do anything save for a consideration, he departed for Dublin, where he stayed till Parliament appointed

I

Trustees to undertake the work at Dagenham, and they advertised for proposals to stop the breach.

A Mr. Boswell tendered for £19,000; but Captain Perry asked £24,000 for the entire work, which Boswell had not included. On this the latter again put in at £16,500! the work to be completed at his own risk, and to begin with his own money! This offer was accepted, and Boswell started on

DAGENHAM MARSHES.

his work. In this he failed. His plans did not serve, and after a while he retired defeated from Dagenham Breaches. The time agreed upon (fifteen months) elapsed, and the opening still defied the contractor, who was told to "stand down," and other assistance was advertised for in 1718.

Captain Perry then submitted his plans in detail, furnishing diagrams, and agreed to go on if allowed £25,000. He immediately began to work, but Boswell sought to undermine him by presenting a petition to Parliament, whither Captain Perry was summoned to be examined on his method, which the petitioners declared to be "impossible." One of the arraigners confessed that he had no personal knowledge of the said methods, nor of the engineer; and after Perry's examination, the engineer was complimented on his observations and mode of working. He left the committee with colours flying, undefeated.

This business had caused considerable delay, but Perry proceeded with his dove-tail pile work. He made other openings in the embankment, so as to reduce the pressure on the big breach, and he proceeded manfully during the year 1716. In the spring of 1717 he finished the second draining sluice, but could not close the dam until June.

This damming of the immense breach had been a hard task, but pile after pile was driven in from opposite sides until they nearly met. These piles were backed and covered with earth, and Captain Perry was most emphatic in his directions to his assistants to use only good earth, not soft stuff, which the "navvies" of the period loved to work in with the proper kind to save themselves trouble. Perry then declared that he would stop the wages of any man using soft earth, and turn all the mixture into the river.

The usual "union tactics" were tried too. When the contractor's need was greatest, the men struck for an advance in wages, though they were earning from 30s. to 36s. a week. Perry was sorely tried, but he got an additional number of hands. Still the work remained unfinished; a tremendous storm arose during the Equinox, the tide rushed over the embankment, and in two hours had carried all the upper walls away, leaving only the dove-tail piles to resist the pressure on the dam.

Fortunately the high tide came in the day-time, and Captain Perry managed to get some small piles driven, with boards put in; so for a while he resisted damage, and certainly prevented it from spreading. He gives us minute details of his proceedings, and bewails his ill-fortune, not as one who has no remedy, but as an unfortunate victim to circumstances. He managed to make his men obedient, and again turned the tide out of the levels by means of his sluices when the high tides did not get over his embankment. He worked hard, raised the dam, put reeds and hurdles upon it, and gradually consolidated it.

But here completion was delayed by the tide again breaking in, when Perry was laid up with ague. The assistants had failed him. The watchman, who had been placed specially to summon assistance if he should observe any danger threatening, had quietly gone away, and left the embankment to itself. These "accidents" crippled Perry; he had to borrow money to finish the work and repair the damage. Emulous workers worried him, but he still pressed on, and in time carried his embankment safely over the level of the highest tide hitherto known.

This was the virtual day of victory. The embankment won! The dam was still strengthened, however, and Perry had the consolation of seeing his efforts fully crowned with success, after five years' hard labour.

Perry lost money if he gained fame. His estimate did not suffice for his needs; his expenditure, as already indicated, did not keep down to the level he had intended. Parliament voted him a sum of £15,000, and another £1,000 was subscribed; but it seems that he was still out of pocket, and "he did not receive a farthing's remuneration for his five years great anxiety and labour."[1]

The incidents narrated above are compiled from the volume published by Perry himself, in which his struggles and misfortunes in the cause of the community are fully set forth. He was an unfortunate man. Compelled to quit Russia because no pay had been sent him for all his time and trouble, he came back to England to be again ruined, after several years of very hard and distressing labour.

[1] Smiles : "Engineers."

FRANCIS EGERTON, Duke of Bridgewater.

BORN 27TH MAY, 1736; DIED 8TH MARCH 1803.

Those who go down to the sea from the London Docks may see the great embankment, and as they pass may remember Perry, and add a stone to his cairn of memory in Dagenham Reach.

And what other romances might be written of the later embankments westward, where evil-smelling Thames flowed slowly under Hungerford Bridge, and pedestrians emerged from the gate into the market. Much might be written concerning the foreshore of the Adelphi; the true "strand" which gave its name to the street now so familiar, the legends and romances connected with Adelphi Dark Arches, and the "Fox under the Hill," the quaint Bargee's "public" on the slimly edge of the tideway, in the spring of the arch, near the Savoy and John Street. Such memoirs scarcely belong to our subject, but the remembrance of the expedition into these somewhat unsavoury nightly solitudes is still vivid, and point to a period before the present northern embankment had been taken in hand by the Board of Works.

OLD HUNGERFORD BRIDGE.

THE FIRST BRIDGE OVER THE THAMES.

CHAPTER III.

OLD CANALS AND LOCKS.—THE DUKE OF BRIDGEWATER AND
HIS FORTUNES.—BRINDLEY THE ENGINEER.

IT is interesting to observe that around Manchester, and also in Lancashire generally, for more than one hundred years, public attention had centred on the inland navigation of England. To the county of Lancaster belongs the honour of originating the first canal in England, and to Manchester enterprise the latest development of the work is due. Not many of our British rivers were navigable without artificial embanking—as already shown in our retrospect of the Thames—and so at a very early period in our history we find the canal or artificial river in consideration.

MAGNA CARTA contains a provision by which all obstructions, such as weirs, are to be swept away "in order that our rivers may be free"; and subsequently by the Act 23, Edward III., it was ordained that "gores, weers, and stakes" shall be pulled up or pulled down or cut, and the sheriffs were commanded to see to it. In later reigns the same tactics were adopted. Henry VI., Henry VIII., Edward VI., and other kings further insisted on these obligations; but even, as in the present day, the riparian owner often assumed a right to close a stream which wandered through his property; and we shall find as we proceed many cases in which landlords objected to the cutting of canals through their grounds.

By degrees corporations were empowered to look after and demand tolls from the users of rivers, to extend the navigation, and to construct locks ("pound locks"). The canal lock is a comparatively modern invention. Some authorities ascribe its first use to the Chinese, who seem to have invented most useful appliances except the steam engine. They had suspension bridges of iron chains hundreds of years before Telford constructed the Menai Bridge, and in embankments we were far behind them. Still, to the Italians, we believe, the canal lock is properly attributed.

At any rate, whether Holland or Italy be the true nursery of the lock, it was apparently the invention of the brothers Domenico of Viterbo, and applied in 1481, and the bar, or barricade, was termed *sostegin*.[1] In the narrative of Telford, who constructed many canals, we find that the practice in England, until a late date, was to "thrust the boat as far as possible to the rapid, and having well fastened her there *to await an increase of water by rain!*"

But when bridges superseded fords and ferries the wooden structure was sometimes built solidly, and furnished with paddles on transverse timbers "opposed to the current." These dams regulated the flow of water, and so, says Mr. Telford, "the arches of old London Bridge were in this sense

[1] "Essays on Engineering, etc.," Earl of Ellesmere.

designated locks," because the arch was cleared of its paddles to permit the passage of a vessel or boat.   It was the custom to close the arches up to low water mark, and by those means retain a certain amount of water for shipping and traffic in the upper reaches.

The "pound lock" was so-called because the water was "impounded" for the reception of the boat ; and so the river or canal was made navigable as at present by these now double-gated locks.[1]   But a rough kind of contrivance for the passage of boats from one level of water to another existed before the "Brothers of Viterbo" improved the inland navigation of Italy.   An Italian writer attributes the "staircase lock" to no less a personage than Leonardo da Vinci, and this seems to have been confirmed by later writers.

We may then be assured that the lock came from Italy ; and in England, in 1755, a company of gentlemen obtained an Act authorising them to make navigable the Sankey brook from the Mersey to St. Helens.   Although this scheme was carried out, and when completed formed the first navigable canal in this country, yet in 1720 Acts had been procured for the navigation of the Irwell, the Mersey, and the Weaver Rivers.   But Yorkshire seems to have anticipated these a little ; though the Sankey was the first true canal cut, the others assisted streams, and made them navigable.   The new navigation was to be "free and open on payment of tenpence a ton tonnage to the undertakers."   This curious clause smacks somewhat of the sister isle—a free canal on payment to the *undertaker* is suggestive of suicide.

This Sankey canal did not "mix with" the brook in any appreciable degree.   It was carried out with three branches and extended eleven and three-quarter miles from the Mersey. There were ten locks.   Coal, slate, lime, and such heavy goods were carried on it.   When these cuttings and new navigations

---

[1] The old Anglo-Saxon word "loc"＝enclosure, has been considered the derivation of the modern "lock."   Dutch is *sluys* ; French, *écluse* ; German, *schleuse* (sluice).

were projected conveyances were scarce. In 1720 there were not more than three or four carriages kept in Manchester, and apropos of this fact we have a curious reminiscence. One of these carriages belonged to Madam D——, of Salford, who being of a sociable disposition, and yet unwilling to conform to the new fashion of drinking tea or coffee, her friends when they came to visit her in the afternoon presented her with a tankard of ale and a pipe of tobacco !

In 1750 hackney carriages were introduced, but in 1710-40 travelling was extremely limited, and goods were forwarded by pack-horses in long single file. Not till 1758 did any one in business in Manchester keep a carriage. The chief want of the time was coal; cheap carriage was necessary, and the canal offered this desideratum. Under these circumstances, and to divert his mind from his unhappy love affair, did the Duke of Bridgewater consult James Brindley.

Each of these men was in his sphere remarkable, and well suited to the other. Both men of decided opinions, they knew each other's worth and respected each other. The Duke, born in the purple, and the millwright, born in the fustian, met on the common level of genius and enterprise. To one as the projector, to the other as the engineer of the most re- markable canal in England must some few pages be devoted. Their work is absorbed in the more modern Manchester Ship Canal, that Aaron-rod of navigation ; but the undertaking will hold its place in the history of enterprise while England re- mains.

Brindley was born at Tunsted, in Derbyshire, in 1716. The Duke of Bridgewater, then fifth son of the first duke, was born in 1736; so there was just twenty years difference in their ages. When the lad succeeded to the title he was scarcely twelve years old, and by that time James Brindley had set up as a millwright and wheelwright—for he had been from a youth devoted to mechanism. While the young engineer had been hard at work,—apprenticed to a millwright at seventeen, and later improving himself—his future partner was cradled in luxury, and in after life led an aimless existence in society and on the race-course.

Lord Trentham was one of his guardians, and Earl Gower, his relative, lived at Trentham, where Brindley had often worked ; for the earl was a great patron of the engineer. Many extracts from Brindley's notes and journals are given by Mr. Smiles, amongst which " Arle Gower " and a " Loog of Daal " figure. The former item indicates the nobleman, the latter the millwright's material ; for Brindley was almost self-educated, and therefore his business entries were generally ill-spelt.

It does not appear that the young Duke and Brindley ever met in any way at Trentham. Their tastes and occupations then must have been diverse indeed. The youthful scion of nobility, who entered so early upon his dukedom, in conse-quence of the death of an elder brother, travelled and amused himself ; while Brindley travailed only. But in his travels Francis " of Bridgewater " had the benefit of the companion-ship of Robert Wood, a cultured man of the world ; and it has been thought possible that the Duke may have gleaned some impressions of canals from his travels in France, and when his mind sought occupation he remembered them.

Notwithstanding all that has been said of the Duke in praise of his canal and his capacity, we venture to doubt whether any such adventure would ever have been undertaken by him if he had not lost his heart and his intended wife. There is no doubt a Providence which shapes our ends, there are combinations of circumstances possible, which compel events as it were. In the career of Brindley we have this combination. Events led the wheelwright to Trentham to cut timber—that was business. Events led the Duke to visit his relatives—that was pleasure ; from the eventual combination evolved the great canal. Before recurring to Brindley's career, we shall close this episode in the life of the Duke of Bridge-water.

A slim built young man, attired in long coat, flapped waistcoat, ruffles, three-cornered hat, knee-breeches, stockings, and buckled shoes, his Grace met, admired, and soon fell in love with the widowed Duchess of Hamilton. This youthful mourner had been known as one of the "beautiful Gunnings." Elizabeth, the youngest, was she who attracted the young duke. This lady had been "wedded with a curtain ring at midnight" to her late husband, who had not lived long after the marriage.

To her Francis proposed, and was accepted, but, even while the wedding preparations were being made, the conduct of the equally beautiful Lady Coventry, the sister of the intended bride, was somewhat loudly called in question. The youthful Duke of Bridgewater evidently considered that not only "Cæsar's wife" but Cæsar's sister-in-law must be above suspicion. He remonstrated, and directed the Duchess of Hamilton not to associate with her sister. This mandate the young lady declined to obey, and when her intended husband threatened to leave her, and cancel his engagement, she, in her proud and spirited manner, bade him go, if he desired to do so!

This quarrel was never healed. The duke was firm, the lady proud. He retired from the world; she plunged again into society, and quickly secured another partner—"Jack Campbell," as Walpole calls him, subsequently Duke of Argyll. The Duke of Bridgewater, only twenty-two years old, retired to his Lancashire estates and busied himself with his private affairs, plunged into business, and thought out the method by which he could bring his coal from the mine to Manchester cheaply, for the public consumption. But he never more made any proposal of marriage. His heart remained seared till he died.

Meantime Brindley, the self-taught, half-ignorant, practical man, was busy. Reared in poverty, his father a spendthrift of such small monies as he had, James Brindley had been apprenticed to Mr. Bennett, a millwright, near Macclesfield, in Cheshire. The lad's energy and aptitude is thus exemplified by Mr. Hughes.

His master was employed in the erection of a paper mill, the first of its kind in that part of the country. The maker fancied he had mastered all the details. Brindley thought otherwise, and the remarks made by a journeyman confirmed the apprentice in his views. The journeyman declared that the mill, as planned, would be a failure. So Brindley set off one Saturday afternoon, walked five-and-twenty miles to the pattern mill, and minutely examined it.

On Sunday he walked back again, and informed his master that the copy was, in some respects, defective. Bennett accepted the corrections in good part, and made the necessary alterations. The result fully justified the action of Brindley. The mill was a success.

The mechanical engineer prospered. He pursued his profession unremittingly; occupied himself with pumping machinery, and drained coal mines. He erected silk mills, and made improvements in the silk-winding machinery. He turned his attention to the steam engine in 1756, but so many obstacles were put in his way by rivals, that in this branch of engineering he did not shine, as in all probability he would have done if jealousy had not thwarted his efforts. Mr. Stuart has pronounced his engine, made of iron, in 1763, the "most noble and complete" piece of ironwork (till then) produced.

We now come to the time in Brindley's life in which his fame was to be assured, and the initiation, working, and completion of the Bridgewater Canal by him and his patron may fittingly have a separate chapter.

JAMES BRINDLEY.

BORN 1716; DIED 27TH SEPTEMBER, 1772.

VIEW IN OLD MANCHESTER—MOZLEY ARMS HOTEL AND BUILDINGS

CHAPTER IV.

OLD COMMUNICATIONS: THE BRIDGEWATER AND THE GRAND TRUNK NAVIGATIONS.

FROM the first part of this volume, and from the remarks which we have made in our last chapter on the state of affairs in Manchester, as set forth by Doctor Aikin, it will have been perceived that communications in those eighteenth-century days were evil, and carriage of merchandise dear, and in bad seasons positively scanty. With no canals to speak of, no roads worth the name, few coaches or waggons, the people in the towns were dependent on pack-men and horses. Coal was a luxury, costing at least three shillings a ton for water-carriage, when so taken by canal for a few miles, and twelve shillings a ton if sent from Liverpool to Manchester by river.

This charge swelled to forty shillings when taken by road, and then the delays were great and annoying.

Even the Duke of Bridgewater, who owned extensive collieries, could not have his coal transported cheaply ; and when we consider that the towing was performed by men we may fully appreciate the tedious nature of the transport.

Brindley's name had reached the ear of the duke, who had, in 1758, applied for an Act of Parliament to construct a canal over, or through, dry land, to span roads and rivers, and to unite his coal mines at Worsley to the town of Manchester ; ten and a half miles over ground, with supplementary workings of no less than thirty miles of tunnels in the mines themselves! In this Act he made important concessions to the public as regarded cost of coal, and in consequence his Bill was backed, and passed easily in March, 1759.

The Duke had in these matters the example of his father, who had had the idea of rendering Worsley brook navigable ; but there is no doubt but that the second duke fully deserved the title of the Father of Inland Navigation. George Stephenson was dubbed Father of the Steam Engine with much less reason in his case. But the Duke plunged into his canal navigations with determination. He cast behind him all his former boon companions and fast associates. Society knew him no more, and people were considerably surprised when the beau buried himself in his Lancashire estates, in company with no one save Brindley, the millwright, and Gilbert, a land surveyor.

John Gilbert had been employed as a mining agent, and had been introduced to the Duke by a relative—probably by Lord Gower, his brother-in-law, as Brindley was introduced by Gilbert. The engineer, we know, had been employed in pumping operations in mines, and nothing is more likely than that the agent should have introduced the engineer to his patron. Brindley then was a very well-known man, and recognised as capable. We must also remember that Gilbert was interested in canals.

This Gilbert seems to have been, in his way, remarkable. A contemporary portrait is preserved of him, in which he is described as "a practical, persevering out-door man. He loved mines and underground works," and by a rash act nearly lost his life in a mine by putting a lighted candle too near the roof, when the "foul air" exploded. He was saved by a collier, who threw him flat down and lay on him in the drift. "The collier was badly burned, but Mr. Gilbert provided for him and his family."

The Duke was attracted to Brindley in consequence of his giving him some hints on puddling. He advised the Duke how to stop a breach in a river, and executed it by means of liquid mud, which has ever since been used. This, then new, method pleased the Duke ; the work was successful, and Brindley shared the councils of his patron and Gilbert. In Brindley's examination before a parliamentary committee, he was requested to explain what "puddling" was. The engineer gave the members a practical illustration of it by moulding a mass of clay into the form of a trough, and pouring water into it. The water ran through. He then worked up the clay with water, as in puddling, and the trough held the liquid poured into it. "Thus," said Brindley, "I form a water-tight trunk to carry water over rivers and valleys whenever they cross the path of the canal."

The three companions surveyed, planned, and acted together in private almost unnoticed. In the manor house of Worsley, overlooking Chat Moss—so soon to be celebrated in the railway annals—the residence of Mr. Bradshaw, they met. There the men plotted, or adjourned to the village alehouse, where they smoked long, in silence, projecting in their minds the future course of the canal, or subsequently, when money ran short, considering how the work could be continued.

From the coal mines at Worsley, the navigation was projected to descend by a long series of locks to the Irwell, and up again in a like series of steps to the same level. But Brindley opposed this plan. His bold conception was to

carry the canal on a level over the river on arches, by the
Barton aqueduct.   No novelty was the aqueduct on the
continent, but we should remember that whatever Brindley
had heard of continental water-ways he could know but little,

THE BARTON VIADUCT OVER THE IRWELL.

inasmuch as he never went abroad, and it is doubtful whether
he could read with ease, if at all.

But the engineer was obstinate.   He carried his point, and
then he was compelled to ride up to London to obtain the

necessary powers for this new undertaking. It was on this visit that he distinguished himself before the committee. When asked with reference to his project, what, then, was the use of rivers, he replied, "to feed navigable canals!"

The Act was passed in 1760, and thus Brindley's ambition was satisfied. The canal was made on an embankment and carried over the river on a bridge, according to the model suggested to the House of Commons committee, who requested Brindley to explain his views of it.

The engineer went out, purchased a cheese, and dividing it into two portions said, "Here is my model!" He then proceeded to show the manner in which one-half of the cheese could span the other, and by whatsoever method he worked. The members requested him to leave his "model" in the room while his examination continued.[1] Perhaps they lunched off it.

The construction of the aqueduct gave rise to many curious speculations. Such a work had never been seen in England, and pictures of the period represent the boats in the river passing beneath the aqueduct, on which are canal boats drawn easily by horses.

A writer in 1763 thus describes the canal and aqueduct :—

" Not long since I viewed the natural curiosities of London, . . . but none of them have given me so much pleasure as in surveying the Duke of Bridgewater's navigation. His projector, the ingenious Mr. Brindley, has indeed made such improvements as are truly astonishing. At Barton Bridge he has erected a navigable canal in the air, for it is as high as tops of trees. Whilst I was surveying it with wonder and delight, four barges passed me in about the space of three minutes, two of them being chained together and dragged by two horses, who went on the terras of the canal, whereon I must own I durst hardly venture to walk, as I almost trembled to see the large river Irwell underneath me, across which this

---

[1] "Memoirs of Brindley."

K

navigation is carried by a bridge, on the 'battlements' of which the horses walked."

Afterwards the subterranean work is mentioned in these terms :—

VIEW OF BARTON AQUEDUCT.

"The navigation begins at the foot of some hills in which the Duke's coals are dug, from whence a canal is cut through rocks which daylight never enters. . . . By this means large boats are hauled to the innermost parts of the hills, and being there filled with coals, are brought out by an easy current.

"At the mouth of the cavern is a water-bellows, being the body of a tree forming a hollow cylinder standing upright. Upon this a wooden basin is fixed in the form of a funnel, which receives a current of water from the higher ground. This water falls into the cylinder and issues out of the bottom of it, but at the same time carries a quantity of air with it, which is received into tin pipes and forced into the innermost recesses of the coal-pits, where it issues out as from a pair of bellows, and rarifies the body of thick air which would otherwise prevent the workmen from subsisting on the spot where the coals are dug."

Some engineers ridiculed the idea of the aqueduct, and went so far as to call the project a castle in the air which no one had hitherto seen. The embankments also were, for the time, stupendous, and carried the canal safely, even on "sideling ground." A stream was carried under the canal level, and many roads were cut down to pass beneath it. "Sometimes the navigation is carried over public roads," writes the correspondent already quoted; "in some places over bogs, but generally by the side of hills, by which means it has a firm, natural bank on one side."

The cost of the canal was one thousand guineas a mile, but money became very scarce with the Duke while it was being constructed. Anecdotes are related in various memoirs how the three "accomplices" and Mr. Bradshaw would puff their pipes and silently ruminate upon ways and means. On one of these somewhat melancholy occasions Brindley cried out,—

"Don't mind, **Duke**; don't be cast down; we are sure to succeed!"

"The Duke," says Mr. Hughes, "devoted the whole of his large fortune to the prosecution of his works, and strictly confined himself to an income of four hundred a year." Nevertheless he was greatly hampered for money, and it was said that the

principal employment of Mr. Gilbert was to ride up and down
the country endeavouring to raise money on promissory notes ;
but "at one time the financial difficulties had become so
great that the Duke could not get cash for his bond for five
hundred pounds !"

Gilbert continued to ride, and endeavoured to obtain subscriptions.   The Earl of Ellesmere relates an interesting episode of this campaign.   On one occasion the agent was joined
by another horseman, and after some conversation the men
agreed to "swop" horses—not while crossing a stream !   On
alighting afterwards at a lonely inn, where Gilbert was a
stranger, he was surprised by the mysterious signs, nods and
winks displayed by the landlord, and indulged in to a degree
which tended to give Gilbert the impression that the man had
lived in Bedlam for a while.

Nor was his mind much calmed by an inquiry regarding his
saddle-bags—whether they were well filled, and an expression
of hope respecting his "luck" !

Puzzled by these and similar hints, the agent revealed himself, and ascertained that though the rider was a stranger, the
horse he had ridden to the inn was well known as the property
of a notorious highwayman ; and the "honest Boniface," no
doubt, thought that the new-comer was a fellow-workman on
the road, to whom civility was the best policy.

Such was the state of affairs ; but the canal was at length
completed.   Gilbert stood by to see the admission of the
water into the navigation, but Brindley was too nervous.   "He
ran away and hid himself at Stretford," says the essayist,.
"while Gilbert remained, cool and collected, to superintend
the operation which was to confirm or confute the clamour
with which the project had been entertained."

It succeeded : crowds flocked to see the canal, and the
Duke was justified.   He brought coal to Manchester cheaply,
and by this canal was the subsequent prosperity of the city
assured.   On the 17th July, 1761, the traffic was initiated, and
success was writ large upon the undertaking.

We need not proceed to describe the Duke's and Brindley's other undertakings. The open or daylight navigation extended from Manchester to Leigh and Runcorn, some eight and thirty miles, the underground canals or tunnels to about forty-two miles. These underground canals were a source of much curiosity, and numerous distinguished men and women visited them.

The Duke of Bridgewater died childless. He never married, and his real estate passed to the Marquis of Stafford, afterwards Duke of Sutherland, by whose descendants the love of engineering is still maintained. The Duke made a large income from his canals, and was perfectly absorbed in them. He had to pay large prices to landowners for the right to run through their properties, but he persevered and succeeded. Every one liked him, eccentric and sometimes rough as he became. He seems to have caught up, by constant association with his men and assistants, a somewhat brusque and rough manner of speech and behaviour. He was never tired of discussing canals—they were his hobby. He loved to travel in his own boats, which brought him a large income by the transport of passengers as well as of merchandise. He died in 1803. He was a plain liver, detested flowers and shrubs; was very fond of smoking and of taking snuff, and of pictures, of which last taste ample evidence remains; but he was no politician. In his later years he became stout, and careless in his dress.

It would surprise people of the present day to learn what Brindley received for his work. No magnificent daily allowances and numerous traffic-passes. He was paid much as a day labourer was paid, and dined for less than a shilling. Steam tugs were tried on the canal about 1797–8, for the Duke would try everything possible; but the paddles of the *Buonaparte* washed away the banks, and the steam vessel was deemed a failure. The Bridgewater navigation is now absorbed in the later enterprise.

We are not attempting to write the life of Brindley, else we might follow him through his other works. He was never idle, and as a rule had too many irons in the fire. His success in partnership with the Duke generated the spirit of emulation in other districts. The " Potteries " began to cry out for a canal by means of which their precious wares could be carried to market ; and with this scheme the name of Wedgwood is

ETRURIA POTTERY WORKS.

indissolubly associated. The Etruria Canal was planned and executed, and its summit-tunnel has made it famous. Our Alpine railways possess summit-tunnels now ; the mountain ranges of Austria, France and Switzerland can show us works which completely dwarf the Pennine tunnel of Staffordshire. But it was the pioneer, and to drive a tunnel a mile and a half

THE MERSEY BETWEEN WARRINGTON AND RUNCORN, WITH BRIDGEWATER CANAL.
VIEW FROM HALTON, CHESHIRE.

long through the hills and carry a canal in it was a feat un-precedented in those days.

Nowhere was communication more needed than in Staffordshire potteries, whose inhabitants were steeped in a morass of misery and vice—poor, uneducated, isolated. Mr. Whitworth treats of the advantages of the inland navigation then dawning, as it were, on England. The way in which fragile pottery-ware was transported on horses was in itself sufficient to plead loudly in favour of reform. It was the old story—Light, more light! Let intelligence and human intercourse penetrate the district, and the people will become cleanly, self-respecting, free and wealthy! The "packman" was the only visitor, the pack-horse the only means of conveyance on those stony ways ; and the Potteries protested, for did not Wedgwood require distribution of his ware?

The Duke of Bridgewater, Earl Gower, Wedgwood, and many other influential people threw themselves into the project, and, of course, met with opposition. The packmen and horse-owners saw that their days were numbered, and even men were found to object to any canal being brought within four miles of a town! The advocates persevered, and the Grand Trunk Canal was constructed from Runcorn by Northwich through the hills into the Potteries to the Haywood junction for the Severn navigation, and also to the Trent valley. It reached "Birmingham by Coleshill," where the iron-workers abounded, and a canal cut to the Severn put Manchester in touch with Bristol in the south, the Channel, Derby, the Humber, and the North Sea.

The tunnel of the Grand Trunk Canal soon became too small for the continually increasing traffic. Many disgraceful scenes of violence and dispute were witnessed at the entrance when some bargemen tried to push others aside. The management did not reflect any credit upon the "company," and after many years had elapsed and public opinion had been heard, Telford, who had devoted himself to the Caledonian Canal, was directed to make another tunnel.

This he did; and so the congestion of traffic was relieved. Telford's tunnel includes a towing-path, which obviates the necessity for "legging," as still practised in many canal tunnels—in the Regent's Canal, for instance. A board called a "wing" is put out on each side, and on those folding-planks the men lie, and kick the barge along.

Brindley was succeeded by Telford, who constructed the famous Caledonian Canal in 1822; but before that numerous other channels had been opened. London and Liverpool were united by the Grand Trunk, the Oxford, the Grand Junction, and Regent's Canals; and from 1777 to 1822 canal engineering became the "rage." Scarcely a district remained unrepresented; trunk lines and branches pushed their way and ramified in all directions. New navigations were pressed on Brindley while he lived, and opened with much display and rejoicing, some not being completed until after his death.

Thus canals became the highways for merchandise, and subsequently for passengers. The accommodation must have been limited and the transit tiresome. We remember the rough and unkempt appearance of the travellers by the Irish Royal Canal in the "fly-boats," which, drawn by three horses, sped along to the west and south, and can recall the feeling of monotony even in the Gotha Canal from Stockholm to Gothenburg by steamer. The railroad companies in after years absorbed many of these navigations, and tried to squeeze them dry by the pressure of competition. But there are numerous consignments which will always pay if sent by canal, and the uses of our water-ways in Europe and America are still many. Who knows but that some day travelling by canal may not become a craze, just as coaching has been in late years! There are advantages in canal travelling which the railroad cannot confer—pure air, and some very enjoyable scenery. And if fine weather be granted, we think a couple of days in a well-equipped barge, sleeping on shore, would pay any company which had the pluck to establish small steam-launches to tow vessels. Well-horsed boats, say on the Grand Junction, would also provide a pleasant outing, and refreshments on board could be arranged for. The "strange adventures" of a canal boat might furnish a novelist with a new subject, and the present writer hereby gives notice of his intention to try the experiment.

A DUTCH CANAL.

A STEAMER IN THE SUEZ CANAL.

CHAPTER V.

THE OVERLAND ROUTE: THE SUEZ, THE PANAMA, AND THE MANCHESTER SHIP CANALS.

IT is not our intention to write a history of the engineering of canals. The result would be dull and monotonous in its entirety. But there are some undertakings, such as the Suez, the Manchester, and the Panama, with some other ship canals, which will repay inspection. The success of the Suez Canal as a financial speculation entitles it to consideration, even if its origin did not place it in the forefront of modern enterprises, as the North Holland Canal and the Caledonian only measure fifty-one and sixty miles respectively, while the Suez Canal is a hundred miles in length, nearly eight times as wide as the first-named, and five times as wide as the latter.

The projector of this splendid work is, as we all know, M.

F. de Lesseps, who had served France in various countries and in divers ways in his time as consul and minister, but had retired from the service to La Chênaie—an historic place, where in country pursuits and occupations he remained planning a scheme which would hand down his name to the latest posterity.

When this writer was a lad there was a very popular entertainment entitled the Overland Route (not the comedy). This was a diorama which was exhibited in Dublin, and which the said writer saw at the time. In this diorama the recent success of Lieutenant Waghorn was delineated by lecture and picture, and a souvenir was distributed, price one shilling, with the description of the route. M. de Lesseps confesses that Waghorn's success had set him thinking of a still better route than the so-called Overland Route, which had little land on its course. To delete this land, and substitute a canal to the Red Sea, was M. de Lesseps'

THE SUEZ CANAL.

idea; but, as he confesses, it was not a new one. Canals had been cut by the Pharaohs, and Napoleon had advocated a passage through the isthmus—but not a ship canal. De Lesseps says that his suggestion was declined "coldly" by the Sultan, and the project slept, because otherwise it would have "interfered with the Viceroy of Egypt."

One morning the news came that the Pacha was dead, and a king who knew Lesseps had arisen. The Frenchman proceeded to the East, and a meeting was arranged. M. de

Lesseps unfolded his views cautiously, and to his delight the Viceroy agreed to the scheme. In his journal the engineer describes his reception, and the manner in which he gained the favour of the Viceroy's suite by his plucky horsemanship. A curious manner of ingratiating himself to the Egyptians— but so it was.

M. de Lesseps had for years studied the problem of the canal. If any one had pronounced it practicable he was met with the reply, "The sand will come in as fast as it is dug out, and faster," recalling the legend of the cleansing of the stable, into which three spadefuls of soil were cast by invisible hands for every one thrown out.[1]

The Pacha looked with favour on the report put before him in a short space. The concession was granted, and the Viceroy said to the Consul-General of America at Cairo, whither the representatives of the various countries had come to congratulate him, "I shall queen the pawn against you ; the Isthmus of Suez will be prized before yours ! "

An excursion across the desert was soon after organized by the enthusiastic engineer, and a party of four, all practical men, started with a regular earavan of camels. Twenty-five were required as water-bearers, and this in a tract of country which has long since been thickly populated.

The immense basin of the Bitter Lakes, then dry salt crust, was explored ; water and food had to be carried. "Not a fly," says M. de Lesseps, "was living in the desert ; and when the party camped at night, the hen-coops were freely opened. No bird, no sheep, would stray away from camp, and even if a fowl lingered in the bush, at the first movements of departure she would quickly fly and flutter up to the camel's back on which her coop was carried."

These are facts ; there is little imagination, but plenty of romantic incident in this most utilitarian scheme. The climate was not pleasing, but amid all difficulties of weather and sand-

[1] "The Black Joke."

storms the volatile Frenchman kept up his own and his companions' spirits. An anecdote will illustrate this.

One of the party had remarked upon the extreme fineness of the grains of sand, and expressed his opinion that they would even penetrate into his watch. De Lesseps advised him to cover himself up when in camp, and the engineer did so, but left a small hole in the covering above his head. After a while M. de Lesseps took a handful of sand and softly let it trickle into the hole down the man's neck and ears !

"See," cried the victim, "the sand has even forced its way through the cloth ! "

He was quite pleased that his theory had been so speedily and wonderfully fulfilled !

The survey was a success, and M. de Lesseps made strong efforts to bring English official minds to understand it. They could not, or, at any rate, would not. Influence at the Porte set up by the " Great Elchi "—our ambassador—tried to strangle the project as "impossible." De Lesseps came to England and saw a publisher with a view to making the work known. "The publisher presented his account for the book," says Lesseps, " in which the largest item is intended for an attack on the work ! " This mode of proceeding astonished while it amused the projector; "for," said the astute publisher, " there is no need to praise a book when it is attacked ; honest people want to see it and judge for themselves."

The report created a sensation, and M. de Lesseps returned to Egypt, while people in Europe considered him somewhat mad, suffering from effusion of canal on the brain. Even the patriotic Viceroy was discouraged by England. But the projector took a tour, in which he learned much, and then undertook a lecturing tour in the United Kingdom, which was successful. De Lesseps' energy seemed likely to triumph over all obstacles. Opposition faded at the Porte, and at length the demand for the money was put forth. The seed had been sown. De Lesseps had travelled thousands of miles yearly, and his harvest seemed likely to ripen.

He called on the Rothschilds, whose terms for floating the eight millions of sterling capital did not please the engineer.

"What will you ask for it?" was De Lesseps' question.

"It is plain that you are not a business man. Five per cent."

"On £8,000,000? That is £400,000! No, thank you; I can do the business myself cheaper!"

France came forward and took all the shares, and the curious display of feeling against the English which induced so many people to subscribe is worth recording. One man only subscribed to be "revenged" on England. Another wanted to subscribe for the "railway in the isle of Sweden," and when it had been carefully explained to him that there was no railway, but a canal; not an island, but an isthmus in question; and not in Sweden anyway, but in Suez, he replied that he didn't care, provided only it was "against the English!"

So France was determined to be "revenged" on her ally, perfidious Albion! But in after years there was a master-stroke played on the other side.

Nevertheless, even with a full share list, the bold engineer found his difficulties not greatly lessened. The Viceroy was carefully absent. He could not, by custom, deny himself to the Frenchman, his friend, but he could absent himself, and he did so, yet he secretly assisted the hero, as we may call him, and did not interfere when De Lesseps took the initiative unauthorized. The Viceroy, though approving, was afraid, and feared for his friend, perhaps needlessly.

There was danger, though, and plenty of it. A police agent, as zealous as that detective who followed Mr. Fogg around the world in eighty days, tracked the trackers, and nearly was the cause of their arrest, even of their deaths. But De Lesseps took the Egyptian bull by the horns; displayed his skill as a marksman, and so impressed the spectators by his slaughter of bottles with a revolver, that, instead of the "anvil" he feared to become, he became the "hammer," and struck while the iron was hot.

The land was taken, and the first sand was turned at Port Said in March, 1859. Immense dredging machines were employed, and carried off 2,763,000 cubic yards of soil a month. There was no great engineering difficulty. All the way of the projected line of canal there is a natural channel dip, and the surface of the desert is hardly higher than the level of the sea. There were certainly some few cuttings required, but the greatest elevation was only fifty-nine feet above sea level.

The work continued, and in 1869, when the Prince and Princess of Wales were in Egypt, the water was admitted to the area of the Bitter Lakes. Later in the year, in December, the actual opening of the canal to traders took place. Even then an unseen obstacle nearly put a stop to the proceedings. A rock which, till fifteen days before the inaugural ceremony, had escaped observation was discovered, and great efforts were made to remove it. Fortunately, they proved successful, and the ceremony was not delayed.

The first passage through the canal was fixed to take place on the 17th of November, and on the previous day Suez and all the stations on the canal were in festive garb. Grandees were arriving from every direction. The Emperor of the French was at Cairo ; the Viceroy of Egypt at Port Said ; and the Pyramids were illuminated. The Emperor of Austria, the Crown Prince of Prussia, both attended, and the latter mingled in the most friendly manner with the representatives of the nation at whose throat his father's soldiers would so soon be flying. England sent a small fleet, and every European nationality was represented on the formal opening ceremony.

All seemed ready. No hitch threatened to mar the success of the inauguration of the Frenchman's grandest work. But suddenly came the report, amid all the roar of cannon and powder compliments, that an Egyptian frigate had gone ashore some twenty miles from Port Said, and was lying across the canal, blocking the passage !

The Viceroy, who was already on his way to receive the guests, hurried back, and every effort was made to dislodge the frigate.   There were three methods, said M. de Lesseps, to be employed : " Either we must bring back the vessel to the middle channel ; fix it on the banks ; or——"

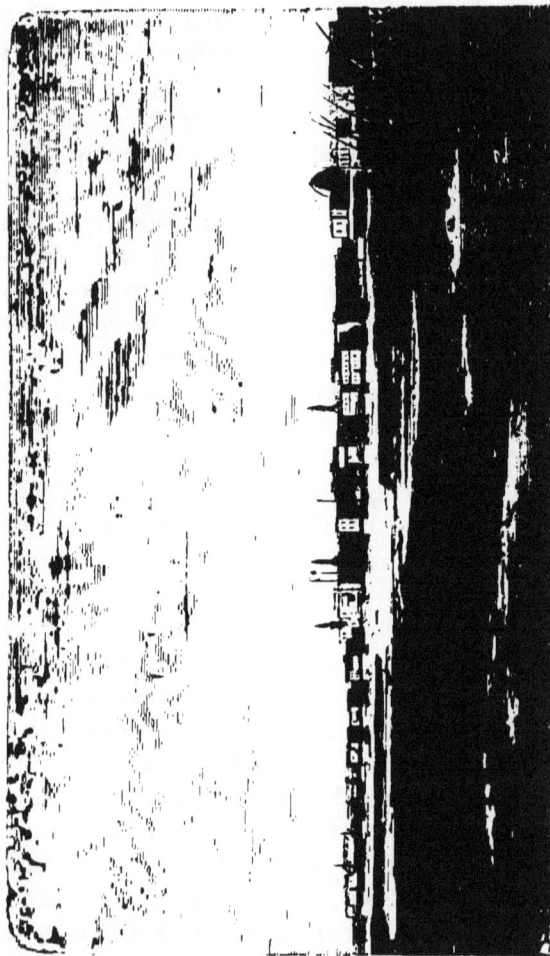

VIEW OF SUEZ.

We looked into each other's eyes.

" Blow it up ! " cried the Viceroy.   " Yes, yes, that's it.   It will be magnificent ! "

" And I embraced him," concludes the engineer. Fortune favoured the brave De Lesseps. The ship was fixed up and saved. Next day, on arriving at Kantara, the *Latif*, dressed in flags, saluted the passing vessels, and "every one was charmed with the attention which had thus placed a large frigate on the passage of the fleet of inauguration ! "

So a failure was turned to a success.

That brilliant Wednesday will not readily pass away from the memories of those now living who witnessed the ceremony. The French *Aigle*, with the Empress, led the way, and the other vessels, Austrian, Prussian, Russian, British, Dutch, Swedish, Italian, American, Egyptian—war-ships, merchant steamers, yachts, gunboats—all passed through from Port Said to Ismailia, the "half-way house," with few hitches or ground-ings, for the ships were so lightened as to draw only thirteen feet of water.

Ismailia was *en fête*, and a splendid ball on Thursday even-ing wound up an important day. The procession continued a desultory way towards Suez, and reached that port in safety. Thus the canal was formally inaugurated, and was subsequently opened for traffic. The whole length is eighty-eight miles ; of this distance sixty-six miles are canal, the remainder lakes ; viz., Menzaleh, Timsah and the Bitter Lakes in that order from Port Said on the Mediterranean to Suez on the Red Sea.

Of the subsequent history of the canal it is not our inten-tion to write. Of the disputes about it, of its widening, of its electric lighting, and the purchase by Lord Beaconsfield of the shares, we need not say anything in these pages. The history will be found in many books of reference. We have briefly traced its course from its initiation to the letting in of the water (in February, 1869, from the Mediterranean side, and from the Red Sea in July), to the grand function which crowned M. de Lesseps with undying fame after his fourteen years of hard labour and almost unceasing toil and worry.

This was the commencement of the type of ship canal of

L

which so many have since been planned from sea to sea, and
from ocean to ocean, to facilitate voyages and quicken communications.   We need only mention the Panama project;
the Nicaragua Canal, the proposed ship canal between Ceylon
and the mainland; the Amsterdam Canal; and last, though
by no means least, our own Manchester Ship Canal.

M. de Lesseps' latest project was stupendous.   In the construction of the Suez Canal he had no engineering difficulties
of any magnitude; hard work easily overcame them.   The level
was easily maintained; no high land intervened between the
now united seas: the Suez Canal was practically level with the
ocean all the way.

But with the Isthmus of Darien the case was different.
Here, however, as at Suez, the idea put forward by the French
engineer was not original.   Cortes and Humboldt were two
of the celebrated explorers who at distant periods had enter·
tained and abandoned the Darien Canal.

In the mind of De Lesseps, Darien disappeared; Panama
arose, and from there to Aspinwall the new canal was to be.
If Suez was a "ship" canal, Panama would be exclusively a
"steamer" canal.   There is no particular industry at Panama,
and the climate is peculiar in the continued absence of wind.
Hence only steamers could navigate the canal, for no sailing
vessel unaided by tugs could hope to enter or proceed.

This peculiarity was a detail for the engineer.   He would
ignore the calms and attack the canal cutting.   Here he hoped
for a level, but was disappointed.   A huge hill met the eyes of
Lieutenant Lucien Napoleon Bonaparte Wyse, whose names
alone seemed sufficient to overcome any difficulties, moral or
physical.   In 1876—his expedition went out in 1878—a concession was granted, the shares were floated by the buoyancy
of M. de Lesseps, and the estimated cost was six hundred
millions of francs.

The work was commenced.   The story is familiar.   Climate
and natural obstacles militated against the workers; things
went wrong; the heat and the stagnation of the place in some

measure affected the promises made by those interested on the spot, and the engagements did not " keep." Thus the works

EARTH DREDGER, PANAMA CANAL.

were interfered with. But notwithstanding all obstacles it progressed, and the mighty Chagres dam began to rear its broad back across the course of the river.

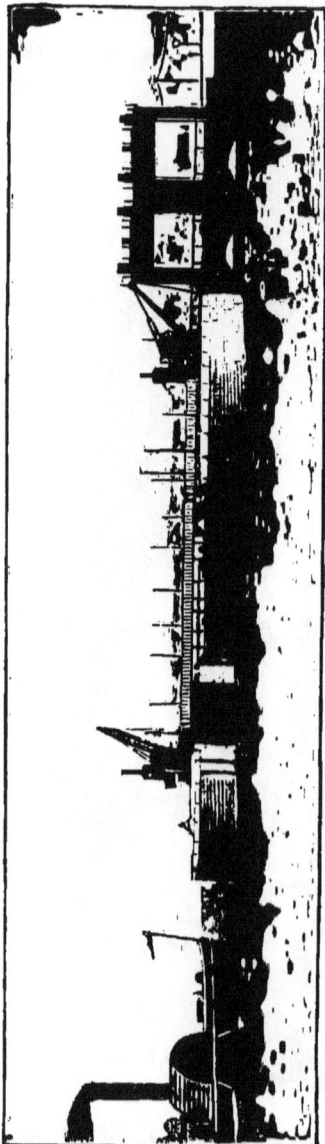

THE MANCHESTER SHIP CANAL: THE EASTHAM LOCK—FROM THE MERSEY.

The Chagres Mountain was, and is, the true obstacle to the progress of the canal. A tunnel was deemed impracticable—a cutting was decided on. A cutting through a mound three hundred feet high was a tremendous undertaking. There was also a river to be taken into consideration—a torrent in rainy seasons, generally in fact, which is capable of rising from a trickle of eighteen inches to a "spate" of forty feet in height, and this without any preliminary preparation or display.

Such a river of such india-rubber-like extension was a factor, and many ways were attempted to meet its idiosyncrasies ; and when it is remembered that this Chagres River was expected to cross the Panama Canal no less than twenty-nine times, the anxieties respecting its playful habits were proportionally great.

The only way was to dam the Chagres, and the dam was begun ; but the work required such immense labour, and money in proportion, that the canal has virtually become bankrupt. Locks were proposed to save the cuttings, and some miles have been completed. But the work is suspended.

The Panama Canal has never been completed, and may never be cut, though arrangements were made in August, 1892, for continuation of the work. The scheme was too great even for the unconquered De Lesseps, who has sunk millions in the attempt to join the oceans.

Of the others named we need say little, save perhaps of the Manchester Ship Canal, the history of which, in its completed form, has yet to be written.

In this project the same energy and determination which animated the Duke of Bridgewater and the French engineer have been conspicuous. When in June, 1882, the party assembled at Didsbury to discuss the practicability of the work some doubts were expressed as to the issue. But though the House of Lords, when appealed to, decided against the scheme, the promoters had faith in the project, and fought the question with such effect that the necessary Act was obtained.

THE EASTHAM LOCK—FROM THE CANAL.

Then the new giant swallowed up the old water-ways. The

famous Bridgewater Navigation and the Mersey and Irwell canals fell into the grasp of the Manchester Canal, which extends from Eastham, on the Mersey, to Cottonopolis. It is nearly completed, and the opening of the water-way on which so much money has been expended will mark an epoch in our history.

The varied scenes of the cutting of the canal have been described so frequently that it would be no new thing to detail the progress of the work. Much has been said of the Steam Navvy, the busy engines and trucks, the villages of huts, the schools, the hospitals, the recreation-rooms on the line of canal ; the steaming, forging, hammering, digging, and building ; the tearing down and turning up ; the locks, the quays, the warehouses and the docks.

Few modern enterprises have been carried out with more care for the workers. Here is no forced labour, no *corvée* ; the navvy—the man and his family—are cared for, body and soul ; taught, and assisted, and amused in "off" hours. That many lives should have been lost was inevitable, but the accidents have not been frequent, and the success of the undertaking is awaiting it, we hope, in the near future.

The employment of the Steam Navvy on the workings of the Manchester Ship Canal gave rise to a great deal of interest and speculation in the minds of those who condemn, or fear, the introduction of machinery for working men. This machine is really a dredger, which by steam power scoops up the soil as a sea-dredger scoops up the mud. Sometimes as much as a thousand tons of "spoil" have been taken off by one of these machines. The cost is great, and the coal consumption considerable, but notwithstanding these expenses there were eighteen of these Steam Navvies at work on the various sections of the canal, representing perhaps £36,000 worth of machinery.

And the human navvy is by no means unrepresented in the panorama of the canal. The invention mentioned does not apparently supersede the genus navvy, who came into existence about the early railway period, and of whom something will be related in our next section. There is little doubt in the minds of the men themselves that, whether the Englishman be the "finest fellow out" or not, the British navvy is the finest workman of his kind. And experience in the Crimea, in France, and everywhere else where the British contractor has planted his army, the English navvy, like the English navy, can "whip" the representatives of other nations. A big feeder, he is a big worker, and few can equal—none surpass—him in his work if he put forth his strength.

And as for independence! He "goes as he pleases," and, then, is not destitute. His mates will always assist him; but he is generally illiterate, and even "the gangers," as a writer points out, "can't tell a B from a bull's foot"—a depth of ignorance for which in these days of education, voluntary, and compulsory, and free, one is not prepared. While the canal is being cut the navigator is sheltered in a wooden hut, the rent of which is mayhap 7s. 6d. to 8s. a week. This house is, of course, let off by the fortunate renter, who will take in lodgers, single men or families, with equal readiness.

Any one who has visited the Manchester Ship Canal has seen these mushroom (not muck-room) villages, with the necessary schools and other "public buildings." The temptation to drink is great, and many men leave their work to go to the beer-shops, which the contractors rigidly exclude from the works.

There is, in consequence, a plan by means of which beer is surreptitiously brought into the sections, and at one time—according to the information of a friend—beer was illegally supplied to the men. To check this policemen made use of stratagem, and in concert, by "simultaneous and harmonious action," as the organ-grinder said, managed to arrest some of the pot-house keepers. On one occasion, it is said, they suddenly appeared from a furniture van, greatly to the surprise of the "villagers" who had helped the van out of the mire, as little suspecting its contents as the Trojans did those of the famous horse which wrought even greater destruction.

The ascent of the canal is by a series of locks of great depth, in sets of three, side by side, for vessels of all kinds.

A guide-book might perhaps profitably be compiled of the Manchester Ship Canal, and of the district through which it passes, and through which the ships will pass when the water-way is opened, as it is hoped, in 1893–4. But there is no denying that the original estimate has been largely exceeded. Just about seven years from present writing the estimated expenditure as mentioned in the Act was £5,750,000. Yet the cost has been much greater even up to date, and it has far exceeded the expenditure thought necessary by the group of promoters who assembled in June, 1882, in Mr. Anderson's house at Didsbury to discuss the great project. When certain lands were included, and the acquisition of the old "Bridgewater Navigation" determined upon, a sum of something over eight millions sterling faced the shareholders.

This, however, as in other great engineering feats, was not the final liability. A series of misfortunes befel the company, Mr. Walker, the contractor, died; there were differences with landowners which delayed the work; and between conservators and engineers much time was lost. To these must be added positive opposition by Liverpudlians, by treaties with Weaver Trustees, and Railway Companies, for completion of the new lines was a *sine quâ non* before the canal could be finished.

The Manchester Corporation came to the rescue, and advanced three additional millions sterling, and on such terms as gave the County Council considerable control in the canal. The Corporation placed five directors on the Board; the three millions were raised on the rates; and, till lately, all seemed well. But now it appears that a further two millions or so will be required to complete the work, and we suppose Parliament will be applied to once more. So about fifteen millions, instead of eight and a half, will have been found necessary to finish the canal.

As to the work performed by the late Mr. Walker and his subordinates, little need be said after we have mentioned that some "forty-eight millions of cubic yards of earth, including eight millions of solid rock," are estimated to be removed from the course of the canal, while the value of the plant employed was valued at nearly £750,000 sterling.

Many interesting relics were found while the canal was being digged, and several old landmarks and associations were removed. Where are the once famous Pomona Gardens, in which entertainments, dancing, public meetings, and private assignations were at one period common? As we pass along the route we recall Ordsall Hall, so intimately associated with "Guy Fawkes," and "Viviana," by Harrison Ainsworth. Where is the Throstle Nest, most romantically named of river locks? And traces few remain of the once picturesque surroundings of Barton-by-Irwell.

From Eastham, where our ship is lifted many feet in the lock, we sail to Latchford, thence to Irlam, and finally arrive at the Old Trafford Docks. The time fixed for the opening of the Manchester Ship Canal is at the end of 1893, and eleven miles of the navigation are now (July, 1892) actually open. This is not quite as originally arranged; 1st January, 1892, was the date upon which the whole canal was to be declared completed.

But in dealing with such unromantic subjects as finance and contract work it is generally necessary to allow a large margin. In the case of the Manchester Canal, the inland city, which wants to be a sea-port, will have to pay for the privilege of "going against Nature." But a great deal of the work is practically completed, and other sections are in an advanced stage and advancing.

A few statistics may close this chapter. The Manchester Ship Canal is 35½ miles long. The average surface width is 172 feet, but this measurement is in places greatly exceeded. Some 20,000 men and lads have been at a time employed on the work, with hundreds of horses, engines, wagons, cranes, and "Steam Navvies." The opening up of this navigation will bring Yorkshire coal to the sea cheaply, and carry goods direct to Manchester, avoiding trans-shipment, delay, and railway charges. These are only some of the advantages of the water-way over the railway and the road.

The first sod was cut at Eastham in November, 1887 ; the contract price was, as stated above, £5,750,000, and over seven millions pounds' worth of shares has been alloted. The arrangement was to finish the canal in four years, but delays have occurred. The changes in the aspect of its neighbourhood have been great, and we trust that the avowed object of the canal, "to afford a cheap means of transit of merchandise of all kinds to and from places beyond the seas," will be amply and profitably carried out.

FERDINAND DE LESSEPS.

FLEET BRIDGE, OLD LONDON (FOOT OF LUDGATE HILL).

CHAPTER VI.

LONDON WATER SUPPLY.—THE ROMANCE OF THE NEW RIVER.

AT a period in our history when the metropolis of England and the metropolis of the Midlands are striving for the first dip into a Welsh lake, and are rivals for its water supply, it will be interesting to look back on the old days when London had no proper means —as at present understood—for the distribution of the first necessity of life—water.

London was formerly supplied by the Thames, the River of the Wells, in Clerkenwell, and by the Wall-brook and Lang-bourne in the city. "In the West," says Stow, "was also 'another great Water' called Old-born, which had its fall into the river of Wells." These Wells, or "fountains," were Holy Well, Clement's Well, and Clark's Well; also Skinner's, Fag's Tode, Loder's, and Reid Wells. All these springs discharged into one channel, which became the River of Wells (afterwards called Turnmill Brook, and finally Fleet Dyke), and was navigable by "ships" as far as the Old Bourne bridge. The Hole-bourne [1] rose from a spot near where the Holborn Bars once stood, according to Stow, but more modern writers say that the River of Wells was the "old burn," or Hole-bourne (in the hollow), and not a separate river.

The running water which entered the city between Bishop's Gate and Moorgate came from the marshy land already referred to, and was called the Wall-brook. It had "divers bridges" over it. The course of this stream is indicated by the thoroughfare to which it gives its name.

Langbourne-Water, so called because of its length, broke out of the ground in Fen Church Street, and ran down Lombard Street and Sherborne Lane ("Sharing" or "Dividing" Lane), where it broke into many rivulets, or rills, and fell into the Thames.

The Wells we have named may still be traced. Smithfield boasted a large pond, and, of course, the Parish Clerk's Well— near where they used to "perform the history of the Holy Scripture"—is perfectly well known. But by degrees all these sources became contaminated and built over, and the citizens were forced "to seek sweet waters abroad"; so one Gilbert Sanford was authorised by Henry III., 1236, to convey water from the town of Tyburn by lead pipes into the city.

This opens up many speculations as to the Tybourne stream and the "town" of Tyburn. We must not go out of our course, though, so we will keep to the water, which fell into a great cistern of lead "castellated with stone," called the Great Conduit, in West Cheap. Conduit Street, no doubt, indicates a portion of the route to the City, from Paddington to James' Head on the hill, to Moorgate and so to the Cross in Cheapside.

Many conduits were built and cisterns established. There were conduits in Aldermanbury and "The Tun" in Cornhill. The Standard was in Fleet Street; at Bishop's Gate and in many other spots were conduits, including one at Holborn

[1] Brook in the hollow.

SIR HUGH MYDDELTON, Knight and Baronet.

BORN ABOUT 1565; DIED 10TH DECEMBER, 1631.

Cross, which was rebuilt in 1577 by Mr. William Lamb—hence Lamb's Conduit and its street. All these conduits were at times visited by the Lord Mayor and Aldermen. Before dinner they hunted the hare, killed her, and returned to dine at the conduit. After dinner, on one occasion at least, they "hunted the fox," and after a "great cry for a mile," the hounds killed master Reynard "at the end of St. Giles's."

Fox-hunting in St. Giles's makes us stare! But in 1562 the Fields extended westward.

Thames water supplied a portion of the citizens, who fetched it themselves; but later some of the lanes were stopped up by the residents, who demanded toll. But the people complained, and the lanes were freed. In later times one Morris, or Morice, a Dutchman, conveyed water by means of pipes to the houses of the citizens.

Morice, the High Dutchman, as Stow calls him, initiated the system of London Water Supply in

CITY CONDUIT.

1582. His plan included water-wheels, which were erected under some of the arches of London Bridge,—there were nineteen arches in those days,—and the force of the tides and currents drove the river water into the pipes with sufficient strength to rise as high as the summit of St. Magnus' steeple, as witnessed by "the Lord Mayor and a goodly companie."

· Such a thing had never been seen, and Morice obtained his charter, or lease of the arches. The first arch on the city

side was first leased for ·500 years at 10*s.* a year; other two
arches were subsequently granted, but they were all disposed
of to a Mr. Soames, who obtained yet another arch, and
turned the undertaking into a company, in 1701. This continued appropriation of arches led to some protest, but as
late as 1767 another arch was fitted up for the water supply
of the Borough on the same principle.

Both Brindley and Smeaton were consulted on this subject,
and the wheels turned merrily, sending supplies to the City
and the Borough through the "mains." This supply in
conjunction with the New River was continued until 1817.
The London Bridge Water-Works Company had determined
to rebuild their wheels, but, says Sir Francis Bolton, this plan
was never carried into effect. The Company was "unable to
compete with the New River Company, which had laid down
iron pipes throughout nearly the whole of the City. The
London Bridge Company, in consequence, transferred all the
leases it held to the New River Company for a sum of £3,750
payable for 260 years." In 1822 the old bridge was taken
down, and rebuilt by John Rennie from his father's designs.
The new London Bridge, a little higher up the river than the
old one, was opened in 1831. The old London Bridge had
existed for six hundred years, and its Company had supplied
the city with water for 240 years!

We must now turn backwards a little and undertake the
rapid survey of the New River, an undertaking whose origin
and progress is too often forgotten in the consideration of the
Company whose shares command so much attention and
curiosity when advertised for sale. This Company is the
oldest in existence in London; it was commenced in 1609,
and inaugurated in 1613.

Sir Hugh Myddelton, whose name is spelt in so many
different ways, came of a Welsh family, and was a noted gold-
smith in London City, as was Soames who had formed the
London Bridge Water-Works Company. In the earlier years
of the rule of James the First efforts were renewed in the

direction of bringing water from Hertfordshire, but no charter was granted. The Corporation would not spend the money necessary; and at length Hugh Myddelton, or Myddleton, determined to do the work and pay for it himself.

On the 21st of April, 1609, Myddelton signed the agreement, and began his marvellous enterprise in May. Telford in after years officially described the situation from which the supply is drawn. "In the valley of the Lea," he says, "in the neighbourhood of Ware, two singularly copious springs issue from the slopes of the chalk hills. The upper and greater is named Chad well, the other the Am well." This reprint is dated 1834.

Myddelton found it no easy matter to unite the wells and lead the flowing water from its pure source down to the Islington Reservoir, some eighty feet above the Thames. The Chad well was tapped, and the

NEW RIVER WATER CRIER.
(*From an old engraving.*)

Am well, which bubbleth up in the heart of the town of Amwell, though inferior in quantity, is superior in quality, was pressed into the service. The actual distance is only about twenty miles from London, but Hugh Myddelton was compelled to carry the aqueduct nearly double the distance to avoid the many natural obstacles.

"The land," writes Howe, "was in some places oozy and muddy, in other places hard and rocky, where he was constrained to cut his trench thirty feet deep ; . . . and at the end of Enfield Valley it is carried with great art over a valley between two high hills." Besides these obstacles, Hugh Myddelton had to resist the opposition of the owners of the land through which his cut was made, and even was attacked in Parliament, where the opponents denounced the undertaking, and sought the repeal of the statutes.

Notwithstanding all opposition and the failure of his funds, Myddelton struggled on, but it soon became evident to him that he could not finish the undertaking within the stipulated four years. The Corporation to whom Myddelton applied granted him another year, and subsequently two more—seven in all. He had 600 men at work, at half a crown a day each.

After a while the projector, rich man as he was, began to perceive that the river would beat him. His funds could not hold out, so he made a petition to the king to assist him. James agreed to pay half the cost if he were given half the shares. To this Myddelton agreed, and the New River was continued, on the understanding that the "interest of the king's share was kept in a subject, not in his majesty." On 2nd of May, 1612, the agreement was made, the king gave the engineer free passage through the royal domain, and assisted Hugh Myddelton greatly.

James immediately paid his share of the money already expended. No further hitches took place, the work progressed fast, and on Michaelmas Day, 1613, the new river was admitted to the "ducking pond," in Islington, which now, much enlarged, is called the "New River Head." There was some ceremony observed at the opening of the river, which had occupied five years in construction, and had cost half a million sterling.

We can, perhaps, imagine the pageant which accompanied the admission of the water, bright, clear, and wholesome, which rushed from its channel into its newly prepared bed.

THE NEW RIVER HEAD (AS IT WAS).

M

Merry Islington that day deserved its name. "On Michaelmas
Day," says the *Chronicler*, "in anno 1613, being the day when
Sir Thomas Myddelton, brother to the said Hugh Myddelton,
was elected Lord Mayor of London, for the year ensuing, in
the afternoone of the same daye, Sir John Swinerton, Knt.,
and Lord Maior of London, accompanied with the said Sir
Thomas, Sir Henry Montague, Knt., Recorder of London, and
many of the worthy aldermen rode to see the cisterne, and the
first issuing of the water thereinto, which was performed in
this manner."

The writer then goes on to describe the troop of labourers,
sixty or more, well apparelled and wearing green Monmouth
caps, all alike carrying spades, shovels, and pick-axes, and
such-like implements of labourers' employment. These men
marched round and round the pool, and then halted before the
Lord Mayor, who, with the aldermen and many other visitors,
stood viewing the procession—a very picturesque gathering,
we may be sure.

The spokesman of the party then stepped forward and
delivered a speech in rhyme, which is too long to quote here,
detailing the work and anxiety the great scheme had caused,
but telling of the courage and hopes of the adventurers, and
bidding others follow their good example. The speech ended
thus :—

> "Now for the fruits then : flow forth gentle spring,
> So long and dearly sought for : and now bring
> Comfort to all that love thee ; loudly sing,
> And with thy chrystal murmurs struck together,
> Bid all thy well-wishers welcome hither."

At the conclusion of this quaint address the floodgates were
opened, and music burst forth in triumphant strains as the
water rushed into the reservoir. People cheered, bells rang
merry peals, the spectators on foot and horseback crowded
to the edge of the pond, and "Hugh Myddelton's Glory" was
assured, as he stood on the left hand of the Chief Magistrate,

who, seated on a white horse, beheld with delight the issue of the New River scheme.

Islington was a full partaker of the happiness thus engendered, and on no former festival were more cheese-cakes and cream consumed ; for

> "Hogsdone, Islington, and Tottenam Court,
> For cakes and cream had then no small resort."

Islington was deserving of the title of Merry in the olden time, when archery and pleasant games and teas in the White Conduit Fields and Conduit House attracted Londoners on pleasure bent. It may not be generally known that a cricket match played in the White Conduit Fields, in May, 1784, led to the formation of the Marylebone Club.

We could wander round the New River Head, and trace its course from place to place, every spot full of interesting associations, if our space were not limited. But the ceremony is over, the New River flows now into London, and its romance is almost forgotten ; the fields are built upon, the river is regarded as a supply. No longer the bathing place for Northerners, it flows sluggishly into its receptacles fed by Hampstead and Highgate and other ponds which are contributaries to the Company's supplies.

We cannot permit Hugh Myddelton and his work to pass away entirely, though. He had brought the water to London, but he had to distribute it ! This was proceeded with with all possible diligence, by means of "pipes of elme and lead."

Many hundreds of pipes were laid, but thousands of people living at a distance from them were daily supplied by water-carriers, who bore pails on a yoke, as milk-men do now, and cried the New River water.

"Fresh water ! New River water ! none of your pipe sludge!" was their cry.

Hone, in reference to this, quotes the lament of the citizens: "Oh dear ! what'll the world come to ? They won't let people live at all, by-and-by. Here they're breaking up the ground,

and we shall all be under water some day or other. . . .
I'll stick to the carrier so long as he has a pailful and I have
a penny."

So the householder preferred the old fashion, and ignored
the Company's supply, which was the cheaper.

In June, 1619, the shareholders were incorporated, and Sir
Giles Mompesson—the original of Sir Giles Overreach—was
appointed surveyor. Subsequently the affairs of the Company
became somewhat embarrassed, and calls were feared, so
Charles I. parted with the king's shares. Sir Hugh, who
had been knighted by James I, was granted the king's shares
for £500 a year, and thus secured the benefit of the other
moiety. This payment is continued by the Company. But
the king's representatives are not entitled to take any part
in the direction of the New River Company, and this in some
measure may diminish the value of those shares. Sir Hugh's
shares are termed "Adventurers'" shares, and are now greatly
split up, as may be ascertained by the perusal of a notice of
sale. It is said that Hugh Myddelton was compelled to part
with many of his original thirty-six, and so his descendants do
not hold such an interest in the Company as they deserve, or
would otherwise be entitled to. In the "Calendar of State
Papers" many interesting particulars of the New River will be
found.

As the great city grew, the demand for water became
greater, and the New River was found insufficient to supply
the citizens. Only twice a week was the water sent into the
houses, so the Lea was accordingly tapped, and the mill-
stream purchased by the Company. But the lowering of the
Lea caused a law-suit, which resulted, after some litigation, in
the Company being permitted to take a certain quantity of
water from the Lea by means of a pipe, the dimensions of
which were specified. "But," says the *European Magazine*,
1814, "the Company found this supply too small, and their
surveyor offered double the price paid for permission to use a
pipe double the size. This specious offer was accepted in all

simplicity, and in consequence the river Lea trustees were compelled to give four times the supply for double the money, because of the mathematical fact that two orifices are to each other as the *squares* of their respective diameters, and the pipe twice as wide gave the supply squared."

It is unnecessary to proceed farther with any statistics. Those who are interested in such details will find them fully and clearly set forth, with those pertaining to other water companies of the metropolis, in Sir F. Bolton's work on the London Water Supply. Particulars of the New River we find in Pink's " History of Clerkenwell," a very interesting survey of the parish and its history, to which we owe our thanks.

There are many curious incidents related in connection with the New River. It seems that King James fell into it, and had a narrow escape from drowning. But in more modern days voluntary bathing was resorted to, and Hone says, " Were the doings in the New River during summer, or one-half of the wholesale nuisances permitted in the Thames described, the inhabitants of London would give up their tea-kettles." The water of the White Conduit was especially condemned.

Sir Hugh Myddelton, " barronett," died on 10th December, 1631, and was buried in the churchyard of St. Matthew, Friday Street. He never received any reward save the empty title bestowed on him by James I. He married twice, and had sixteen children. For years no memorial was erected to him—only a sign of a public-house commemorated the self-denying citizen ; but at length a monument was erected at Amwell, on which suitable inscriptions were cut. A statue and drinking fountain were also erected, in 1862, at Islington—the water is supplied gratuitously by the New River Company ; and so the memory of Sir Hugh Myddelton is preserved near the spot which witnessed his " glory " in 1613.

> " Amwell, perpetual be thy stream,
>   And ne'er thy spring be less,
>   Which thousands drink who never dream
>   Whence flows the boon they bless."

The following extract from one of the London daily papers will indicate the value of the New River holdings in this year of grace, 1892 :—

"THE NEW RIVER COMPANY.

" An adventurers' share in the New River Company was submitted for sale by auction, at the Mart, Tokenhouse-yard, by Messrs. Edwin Fox and Bousfield.   The possession of this freehold share in the Company confers on the owner several privileges peculiarly associated with this historic corporation. The share produces a present income exceeding £2,600 per annum, and practically qualifies the purchaser for a seat, with its emolument, at the Board. For this investment it is claimed that it commands an advantage over Government Stock in having a progressive income, besides the prospect of an annually increasing revenue and a valuable reversion, in twenty years, to landed estates of considerable worth.   The annual revenue of the Company—£532,754—which is nearly double what it was twenty years ago, is derived, not only from water rents and water sold in bulk, but from rentals of landed pro- perties in the counties of Middlesex and Hertford, and in the City of London.   Only one of the adventurers' shares has been sold by auction before, and that was in 1889, when it was pur- chased by the Prudential Insurance Company for £122,800.

The room at the Mart was crowded when, shortly after two o'clock, Mr. Bousfield began the auction by a statement of particulars.   He pointed out that the share consisted of three several parts of adventurers' shares—viz., one-half, one-sixth, and one-third—and has been so held since the year 1740. He intimated that an offer had been made for it of £110,000, but refused, because the sale of the share was to be by auction ; and that as there was no difference in the value since the last one was purchased three years ago, except so far as it might be affected by the increased income of the Company, he sug- gested an advance in the price upon that obtained in 1889 as being a reasonable hope under the circumstances.   The bidding opened very slowly, and the intervals were filled by the auctioneer meeting objections that might have been raised, amongst them being scares concerning the London County Council, the possible buying up of water companies' interests, and the supply of water from remote sources. The first bid was £80,000, which was advanced, at intervals, by £5,000 at a time to £95,000. Then the offers were made by single thousands till £99,000 was reached. The caution increased, each bid adding £500 only. When the figure attained was £100,500, a hundred was added at every bid rapidly up to £100,900, and, with occasional pauses, from that sum to £102,000 to 103,500. Then the offers became bolder, being £500 at a time, until they amounted to £105,000. Single hundreds increased the figure, with scarcely a breath between each, to £106,000, at which sum the share was knocked down. The name of the purchaser was not made known."

THE VALE OF HEALTH, HAMPSTEAD.

THE DAM—LAKE VYRNWY.

CHAPTER VII.

THE LIVERPOOL WATER SCHEME.—A VANISHED VILLAGE.—THE
STORY OF THE SUBMERGENCE.

F canals and waterworks it may be said that " of
their making, as of the making of books, there is
no end." But, as a matter of fact, of this present
volume there must be an end; and, therefore, it is
within these pages impossible to set forth in order the many
interesting projects which could be treated of in a full romance
of the water-way. We have seen how London was originally
supplied with water; we know, by experience, how it is at
present supplied by companies who obtain it, and, as it is
said, filter it in large beds, before they distribute the precious
—not to say dear—fluid to our houses.

The romance is pretty well crushed out of it, or, shall we

say, pumped out of it, by this time. The process by which our large towns are supplied is practical and profitable. But sometimes in even these prosaic days a gleam of romance penetrates into the hard array of commercial details, and illumines the enterprise of the business man and the engineer.

Such an oasis in the dry desert of facts and figures is afforded to us in the examination of the project of the Liverpool Water Supply, which is obtained, as every one is aware, from a Welsh valley, as the supply of Manchester is drawn from an English lake, and that of Glasgow from Loch Katrine.

Some years ago Liverpool cast an eye upon Bala Lake, and proposed to tap it for the needful supply which the rapidly increasing population of the city demanded. There were wells and reservoirs, but these only served in part, in consequence of leakage, which was rectified. All seemed safe, but Liverpool would go on growing, and "more water" was the cry. Several English lakes, amongst them romantic Ulleswater, were suggested as fit for the honour that Liverpool citizens proposed to confer on it. Ulleswater should have the freedom of the city, said the Corporation, and "the run of the streets"; but the proposal was not accepted.

The successive disappointments might have wearied many men, but your engineer, who has a good berth, is not easily discouraged; nor indeed is he a man likely at any time to pronounce the word "fail." So when the engineer of the Liverpool Water Scheme found that he could not tap the Welsh or the English lakes, he hit on a brilliant idea, and elected to draw his water supply to Liverpool from the dry land!

How he proposed to do this and how he succeeded we shall shortly relate.

The site on which his experienced eyes had rested is a pleasant valley, nearly eighty miles from the city, in which the village of Llanwddyn was situated; was is used advisedly, the village has disappeared! There is a touching side to this picture, which we studied in 1889, when on a course of engi-

THE VYRNWY VALLEY.

neering. In that valley of Montgomeryshire, sheltered by the Berowin Mountains, lay a few years ago a smiling village, free from the busy hum of men, but not far from the madding crowd of tourists who visit the Arran hills and Dolgelly, or Bala; amid a perfect flock of "Llans." From Bala two roads lead in a south-easterly direction by the Vyrnwy River, and near the village the roads united.

This valley, almost enclosed by hills, is like most of North Wales—rainy. The reputation for moisture enjoyed by the district is not so appreciated by the tourist, whose lamentations, copied from an hotel book, will explain the situation :—

> "The weather depends on the moon as a rule,
>   And I've found that the saying is true,
> For at Bala it rains when the moon's at the full,
>   And it rains when the moon's at the new !
>
> "When the moon's at the quarter then down comes the rain,
>   At the half it's no better, I ween;
> When the moon's at three-quarters it's at it again,
>   And it rains besides, mostly between." [1]

In such a district, once covered with the glaciers of an ice age, which drained into the valley and formed a lake long dried up, the Liverpool engineers, (the late) Mr. Bateman and Mr. Deacon, perceived their opportunities. There was plenty of water in the Vyrnwy ; the mountain torrents poured many tons of water down the mountains ; why not catch this unlimited supply in the valley, make a strong dam, and turn the whole glen into a reservoir lake ?

A bold project ! The village, the church, the churchyard, the cottage and the hall, the shops and the post-office, what is to become of them ? What is to become of the inhabitants ? and, above all, what is to become of their beloved dead, sleeping their last sleep in the quiet "God's acre" yonder, within the shadow of the little church beside the road, beyond which rise the undulating sandstone hills, whose scarped and precipi-

---

[1] See Mr. Roberts' interesting "Gossipping Guide to Wales."

tous sides tell of ice and moving glacier? What is to be
done with all these good people, alive and dead?   Evict

ROAD OVER THE DAM, LAKE VYRNWY.

them ! Turn them out from their rest, from their houses,
and what some believed were their last homes !   Let not
sentiment interfere ; " Nothing is sacred to the sapper !"

The threatened overflow of the waters was told to the people. There was no ark to save them, neither would their valley be any more dry land when once the rains were diverted to the hollow. By degrees there rose a mighty dam across the narrow opening of the valley. The Act of Parliament had decreed the obliteration of the village, *Sic volo, sic jubeo!* The "rights" were purchased; dead and living were removed from vault and grave, from house and home. The young and the old migrated, with their hallowed dead, and then the dam was shut, and the windows of heaven opened.

We can picture the scene; and the tourist will find his trouble well repaid by a visit. The beauty of the Berowins, or Berwyns, is unchanged. Around the now artificial lake, in the valley, once a natural lake in which prehistoric beasts wallowed and tumbled in their monstrous play, a road can be taken and the scene viewed. The visitor may see, perchance, the chimneys and the roofs "in the waves beneath him shining," and meditate upon the curious fate of Llanwddyn.

Inch by inch the water increased and filled up the ample bosom of the glen. The still lake, the wild surroundings are impressive. The great dam, commenced in 1881, is 161 feet high, 1,172 feet in length, and 120 feet thick at the base. Founded, literally, upon a rock, or rocks, by which, ere Nature burst the barrier, the old time lake had been confined within the valley, the dam remains cemented, a work which, in the ordinary course of events, should last for ever.

At the end of this extensive lake, whose waters press the dam with a force of 167,000 tons, rises a Straining Tower, through which the rainwater is percolated, or strained, into the aqueduct which conveys the supply to Liverpool. This is a picturesque structure, and it is a curious fact that the useful is in this instance combined with the beautiful. The tower is connected with the shore by a little bridge. The water supplied is taken in below the surface of the lake, so that the chances of impurities becoming mingled with the supply is lessened; and this water is again passed through a screen of copper gauze.

Any risks of the choking of the gauze is carefully obviated by a most ingenious but very simple process. When the gauze begins to be clogged and unable to perform its duties satisfactorily, the intake rings a bell, and summons an attendant, who shuts off the supply, raises the screen by hydraulic pressure, cleans the strainer, and sends the waste off by a refuse-pipe. Thus the constant supply is obtained, and there is no lack of water. The lake is five miles long, half a mile wide, and nearly as deep as the dam. The number of inches of rain which fall in the district is considerable in the year. One "inch of rain" upon an acre is equal to 22,622 gallons, or 100 tons of water on an acre. As the average rain-fall in this kingdom is forty inches, and in North Wales and the English Lake districts probably sixty to eighty inches, the amount of water which is collected for Liverpool may be calculated by those who are fond of figures.

The aqueduct is also a fine piece of engineering. The water is conveyed by Oswestry and Malpas to Runcorn. There it crosses the Manchester Ship Canal and the river Mersey. But it is again stored within the immense Prescot reservoirs, whence it is drawn off as required by the Corporation.

The "fall" of the culvert is about seven feet to the mile, but the flow is not uninterrupted. At intervals along this long course of sixty-eight miles or so there are filtering beds on the principle originally adopted by Mr. Simpson, which includes the use of fine sand, and, beneath, shingle or gravel, in proportion to its coarseness, undermost. There are in the course of this main-pipe several tunnels, and it manages to pass over or under railways, canals, rivers and streams, in tubes, and culverts, and pipes.

Such is the somewhat romantic side of the Liverpool Water System. Gathered from earth—no one knows where—the rain falls upon the hills, is carried down from Welsh mountains to Liverpool streets, to be again absorbed; and, for all that man can tell, re-carried through the same processes, in a scarce vary-ing round throughout the centuries! There is no decay in

Nature not provided for. The same material serves in different conditions; and then the question rises into speech, "What becomes of Man when he passes away into clay? What has become of the flesh and muscle of the dead who were laid in Llanwddyn churchyard, when their bones were removed from it?" "What has become of last winter's snow?" asks the French poet. What cycle of change does the drop of water of the Vyrnwy Lake pass through from the time of its reception amid all the other millions of drops in the lake until it is once more poured with them upon the earth to do its duty in a purer state?

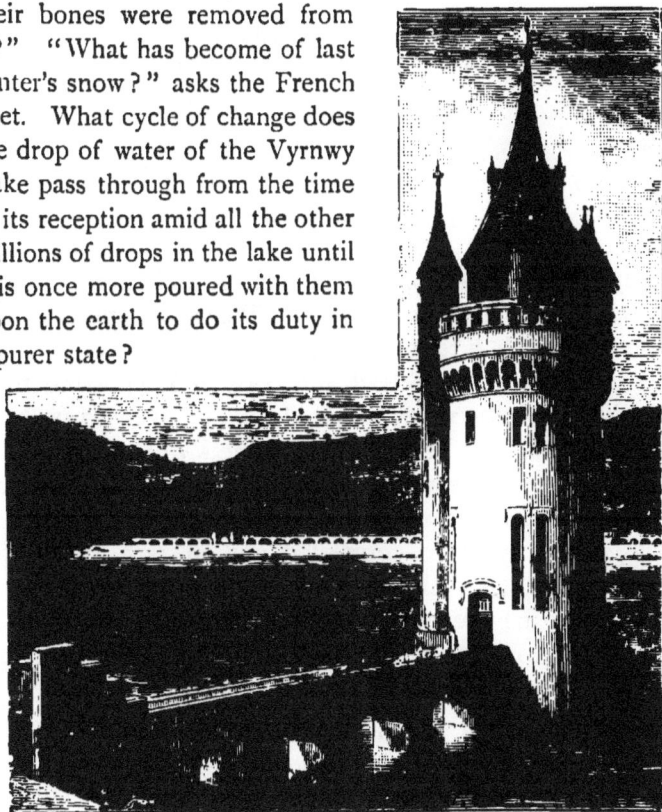

STRAINING TOWER, LAKE VYRNWY.

We wonder whether, when we leave this sphere, shall we ever return to it purified, useful, intelligent, even as we left it, but improved. We may not peep from the lower to the higher life, but Nature never changes her methods. All life tells us of decay and re-creation in other forms or types, of forms like and unlike.   Is Man, then, the only created thing which shall no more re-appear on earth after going through the mill of change and the water or fire of purification ?

But to return to our valley.   Of it may be truly said,—

> " Here rolls the deep where grew the tree.
>   O Earth ! what changes hast thou seen ! "

Changes indeed !   Roads, houses, landmarks, buried beneath the waves as completely as if some convulsion of Nature had swept away the village and the vale—" utterly swept away, as Thebes, or Babylon, or the land of Lyonesse."

There it is ; there the houses dismantled will remain ; the walls upstanding sheltering the finny tribe and telling in their own way.of the destruction of the Welsh village, of *malice prepense*, to supply a town with water !

Yet there is balm in Gilead and corn in Egypt—the town is supplied ; but the valley is not dead, nor likely to be.   An immense interest has been aroused in the Vyrnwy district, and it is a tourist resort as well as a watering-place.   The destruction has in a measure extended its popularity.   The aqueduct and reservoirs have been completed, and on 14th July, 1892, the water from the Welsh hills was loosed by the Duke of Connaught in the streets of Liverpool, having travelled in all seventy-seven miles from the lake.   The aqueduct from the valley to Prescot reservoir is sixty-eight miles, a work of which even Imperial Rome, or Republican America, where " they do big things," might be justly proud.   The Claudian aqueduct is as much shorter than the Deacon pipe-line as the latter is shorter than the Manchester and Thirlmere aqueduct.

The pipes are buried in tunnels or led over arches according to the configuration of the country through which the water is carried.   The great Vyrnwy dam and this pipe line rank among the marvels of the age.

THE FIRST TRAIN ON THE STOCKTON AND DARLINGTON RAILWAY,
1825.

THE ROMANCE OF THE RAILWAY.

CHAPTER I.

ROADS AND RAILROADS.—THOMAS GRAY.—CANAL MONOPOLY.—
THE BIRTH OF THE RAILWAY.

THOSE who have had patience to read the foregoing
sections of this volume will have noticed how locomotion has played the principal part in our book.
The means of getting from place to place, the love
of travel, in fact, is firmly implanted in the minds of the present generation. The social advancement of this country, or

any country, is traceable by its means of communication.   We
have seen the opening up of roads and canals, the development
of communication, the increase in trade, and the spread of intercourse and civilization, in consequence of such facilities of
communication. We have now to look at the greatest triumph
of the whole—the initiation and spread of the modern railroad,
which almost leaped, Minerva-like, fully equipped from the
brains of the miners and their engineers.

It will be in the recollection of many readers that the Duke
of Bridgewater had his fears concerning the birth of the railroad, or the increasing use of the tramroad. When questioned
upon the financial success of his canals, he is said to have
replied,—

"Oh, yes ; they will last my time, but I do not like the look
of those tramroads;" and he confirmed his opinion with a term
of distinct condemnation to emphasize his feelings toward
them.

These tramroads were then in use, and long before the
duke's time we read of the employment of such means. Thus
Roger North, the ancestor of the present youthful Lord Guil-
ford, writes:—

"Men have pieces of ground between the colliery and the
river. They sell leave to *lead* coals over the ground. . . . The
manner of the carriage is by laying rails of timber from the
colliery down to the river, exactly straight and parallel, . . .
whereby the carriage is so easy that one horse will draw four
or five chaldrons of coal."

This was written in 1676. So the "tram" road, whatever its
derivation—and writers as well as speakers differ—very quickly
became popular in the north.   These were only wooden tram-
ways, though—"wayleaves," as the Lord Keeper North terms
them—made of beech, and fastened to wooden sleepers.   On
these primitive tracks the miners' waggons were run alongside
the Tyne to the platform, or "staith," from which the coals
were shot into the barge or collier.

These wooden roads must have quickly been worn out by

the continual friction of the "wayne" or waggon wheels, so that a new layer of surface speedily became necessary. Thin plates of wrought iron were subsequently laid upon the timbers, and this improvement is attributed to the Coalbrookdale Company, to whose enterprise we are indebted, as already stated, for the first iron bridge in England.

The following extract from the transactions of the Highland Society fixes the date of the first step in the direction of improvement, a date anterior to the adoption of the iron plate by the Coalbrookdale Company—viz., 1738 :—

"In 1738 cast-iron rails were first substituted for wooden ones, but owing to the old waggons continuing to be employed they did not succeed on the first attempt. However, about 1768, a simple contrivance was attempted, which was to make a number of smaller waggons and link them together; and by thus diffusing the weight of one large waggon into many the principal cause of the failure in the first instance was removed, because the weight was more divided upon the iron."

The iron rail was again improved by a flange, called a "trammel," which retained the wheels of the waggons in their straight course, and thenceforward the lines became "trammel roads." With our taste for cutting every term as closely as possible, the word quickly became "tram"-way or tram-road, and the year 1767 saw the initiation of the improvement. Some writers have accounted for the "tram"-way by the abbreviation of the name of a Mr. Outram, who at one time was actively engaged in this means of locomotion; but we fancy the trammel road is the earlier and likelier origin of the present popular term.

Mr. Robert Stephenson claimed the first use of the iron railroad for the Coalbrookdale Company in 1767, but Mr. Curr says he was the inventor: "The making and use of iron railroads were the first of my inventions, and were introduced at the Sheffield Colliery about twenty-one years ago." This was written in his "Coal Viewer" in 1797. So that would fix the date of the introduction of the iron rail at the year 1776, practically the same period, as the writer says "about" twenty-one years previously to 1797.

The use of iron for wheels was introduced in 1753, but was not much adopted. In 1800 Mr. Outram laid his railway with stone props in Derbyshire, again following the Newcastle method of Mr. Barnes, who had employed stone supports, or "chairs," three years previously. It is curious to note that the fall in the price of iron-"pigs" should have led to the

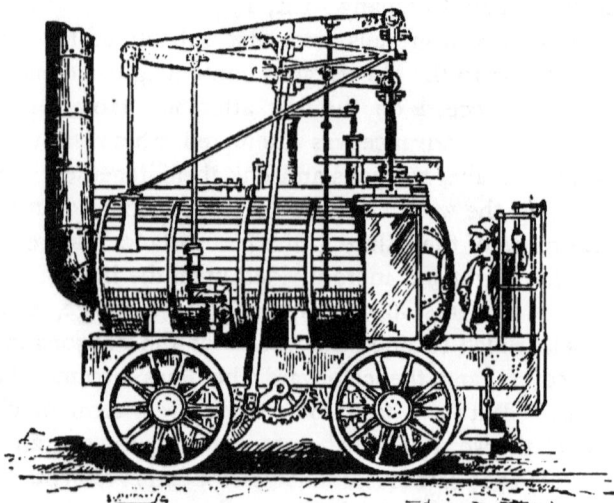

"PUFFING BILLY," 1813.

adoption of iron for rails ; but even then inconvenience was felt in the inclined planes, and the plan of making loaded waggons draw up the "empties" was adopted.

The first railroads, then, were of wood, and the waggons were drawn by horses. The adoption of iron wheels to iron roads or ways increased the adhesive force, while the resistance was very small. Hence the adoption of the "tram" or railroad. But it is extraordinary to the present generation to read that its application to the transport of passengers was so long delayed.

The "improvement" in rails and wheels continued. The tires were made slightly grooved, so as to fit the rail more closely; but this plan was dispensed with when the natural results followed—the fit became too tight by wear, the friction was too great, and so the flat rail was laid.

In 1801 the "railway" came south. In that year the Surrey Iron Company had the honour of constructing the first "railroad" from Wandsworth to Croydon, and in the course of experiments carried men upon their waggons. Here was the germ of the future railroad as we understand it—a means of locomotion for men and merchandise.

SYMINGTON'S ROAD STEAM-CARRIAGE.

This line was, however, called a tramroad; but the development of the powers of steam in this way attracted considerable attention. The pumping-engine was in use; Watt had made many important discoveries; other engineers were turning the idea of the power of steam in other directions. Watt had foreshadowed the steam-locomotive, a steam carriage was made, and in 1802 Trevithick—that unlucky, unhappy genius—actually made an engine to run on the tramroad at Merthyr Tydvil. Thomas Gray and Dr. Anderson were both fully impressed by the usefulness of the railroad, if it could be adopted even for horse-traffic, where canals could not be made.

This gentleman was extremely enthusiastic. "Diminish carriage expenses," he cried; "diminish carriage expenses but one farthing, and you widen the circle; you form, as it were, a new creation, not only of stones and earth, but of men also; and what is more, of industry, happiness, and joy."

The saving in the adoption of the rail was distinctly shown. Thomas Gray, the commercial traveller, saw the trains of coal waggons, and exclaimed, "Why are not these roads laid down all over England, and steam-engines employed to convey goods and passengers along them, to supersede horse-power?"

The reply he received is indicative of the prejudices then existing. His friend said,—

"Propose that to the nation, and see what you will get by it. Why, sir, you will be worried to death for your pains."

But Gray was not discouraged. He saw all around him evidences of the efforts of commerce to burst its bonds. It was struggling to extend its channels, and even on the line of the Bridgewater Canal the block of merchandise had become intolerable. The burden of monopoly was greater than the trade could bear. Machinery had by this time come to a union with steam, and production had increased in consequence. Arkwright, Hargreaves, Crompton, and other operatives had given an impetus to cotton manufacture, and the engineer must find means of transport, which the canal owners made such charges for, and were slow then.

These canal capitalists prepared their own downfall. The manner in which the companies made a favour of transporting goods, and the manner in which they handled them would put the most reckless "muddle-puddle porter" to shame. Managers held autocratic sway, and actually kept crowds waiting their pleasure—the managers' pleasure not the crowds'—while they decided whether they would, or would not, take certain freights. They abused their authority more than even a water-company dare do now-a-days, and by such means undermined their business, as such monopolists always will do.

When the transport of goods from Liverpool to Manchester

occupied a longer period than the then passage from the port to New York, the worm-like public began to turn. Detention in warehouses, from want of barges, or stress of business, or stress of weather, or caprice of manager, or "bargee," or want of water, closing of route, and various other reasons, had caused a tremendous accumulation of material; but the canal people did not care if mills stood still and "hands" stood idle and hungry! It was not the canal manager who cared! His company would only carry timber—and then only one kind! Go where else you please; don't bring your bales here!

In despair, the shipper tried another of the transport companies, who lived and fattened on the tolls of carriage, and yet limited it as much as they could! But he was met in the same spirit of obstruction, and if he did not beg he had no chance. "We will carry it or not, as we please, at our own time, take our time, too; and if you don't like our terms and charges, try elsewhere." This was virtually the argument of the Canal Company.

So commodities literally "stank in the nostrils" of the people. The consigners were fined because they did not distribute the goods consigned, which lay in bulk, and in baulk, in the streets, or quays. Such statements were made in Parliament as we in this age of rapid distribution can hardly credit, and we have not in any manner exaggerated them. The inconvenience thus caused made people reflect, and to them came Thomas Gray, who published his "observations"; but his pleas were called crotchets; his views were regarded as merely "visionary!" He pressed his subject home on all occasions, and, according to Hewitt, "enveloped you in steam" on the slightest provocation.

Thomas Gray, of Nottingham, must certainly be regarded as the pioneer of the railroad. There have been rival claims set up on behalf of a Mr. James, of London, and on behalf of a German named Friederichs; and there is testimony on behalf of each claimant which we do not intend to weigh in full on either side. The general conclusion of writers of the past seems to point to Thomas Gray as the warmest and most original advocate and designer of the rail-road, and with this we may rest contented, but while resting, may read Mr. Wilson's testimony.

On the return from England of this old friend of Gray to Brussels, where both resided in 1818, Wilson visited Gray, but found that no one was permitted to enter his apartments. It was a "sealed book," as he puts it; and the student's brother informed the visitor that Thomas "had some mysterious work in hand which could not be named!"

Subsequently, the "inventor" informed Wilson that he was on the verge of completing his labours : work which had occupied him for years, which would revolutionize society. Mrs. Gray, when spoken to on the subject, merely opined that "the writings" would "make her husband ill again," and she continued, "I ask you what good will it do him that, as he maintains, he is busying himself with the happiness of human kind!"

Evidently Mrs. Thomas Gray did not to the full appreciate her husband's unselfish devotion to the interests of the human race.

The project when discovered was nothing less than "Observations on a Railroad for the whole of Europe." "Here," exclaims the author of it, "is the mainspring of the civilization of the world; all distances shall disappear; people shall come here from all parts of the continent without danger and without fatigue; distances will be reduced one-half; companies will be formed; . . . the system shall extend over all countries. . . . The discovery will be on a par with that of printing!"

This volume, entitled "Observations on a General Iron Railway, or Land Steam Conveyance, to Supersede the Necessity for Horses in all Public Vehicles," was published in London in 1820, and Thomas Gray was condemned as a madman. The *Edinburgh Review* distinguished itself by suggesting that

Gray should be shut up in Bedlam; the same *Review* predicted that no steam ship would ever cross the Atlantic !

Gray seems to have gained nothing by his efforts; Mr.
James, his rival claimant, went to Mr. Sandars, of Liverpool,
and offered to survey the suggested Manchester and Liverpool
Line. Mr. Sandars agreed, paid the expenses, and the survey

GEORGE STEPHENSON.

was made in 1822; but the plan of the line was for the time
permitted to drop.

Meanwhile, an association of the Society of Friends in the
county of Durham had already projected a tramway; and
had, after some difficulty, obtained their Bill in 1821. This
Act was "for the passage of waggons and other carriages

from Stockton to Witton Park Colliery, Darlington." It was originally projected as a tram-road, and it could be used by the public with cattle and carriages during stated hours. The projector of this line was Mr. Edward Pease, who lived to the ripe age of ninety-two, and saw the advantages claimed for the railway system fully realized.

Upon Mr. Pease called George Stephenson, and Nicholas Wood, a writer of a treatise on railroads, then a viewer at Killingworth Colliery. George Stephenson explained his errand and his views; he saw his way to certain developments of the railway system, and advocated an iron road of rails, not a tramway. Not only that, but he made Mr. Pease stare by advocating the employment of steam machines—locomotive engines—in the traction—not horses at all!

After some conversation the visitors left, and the scheme was floated. Stephenson and Wood walked back on their road to Newcastle as far as Durham, while Mr. Pease pondered. No one would credit the locomotive. A horse was much better. Lord Eldon generously, but very foolishly, decided to "eat all the coals your railway will carry!" Every landowner resisted the railways. Agitators told them that no one would need horses soon, and the farmers would be ruined!

The Stockton and Darlington "Quaker's Line" succeeded in passing Parliament; George Stephenson was engineer, and his engines were employed. On Tuesday, 27th September, 1825, the line was opened. The scene has been put before us by Mr. Pease himself in his speech during the Railway Jubilee. He said that the scene baffled description. "Many who were to take part in the event could not sleep during the previous night," and rose at midnight! People cheered or looked nervous, according to temperament. The procession moved off; the train moved on, preceded by a horseman! Fancy the "Flying Scotsman" preceded by a man on horseback with a flag! The rider would have Black Care behind him very quickly in the shape of a "single" eight-feet driving wheel engine, if he tried the experiment now.

But on that auspicious day the train went at eight miles an hour. " No. 1 " was driven by Stephenson : it pulled six loaded waggons, a passenger "coach," and twenty-one coal waggons crammed with passengers ; after which came six coal-trucks filled with their usual mineral contents. A very considerable load for " No. 1."

The excitement was intense, as may be surmised. On went the pilot, after him the train. All along the road and embankments were crowds, rushing, running, riding, cheering, galloping along, in sight of the train, till Stephenson, telling the horse-pilot to get out of the road, put on steam, and soon left the excited multitudes panting in the rear. There was only one sorrowing heart in the district, and that was Edward Pease's.

The worthy Quaker had been singled out by Providence on the day of his triumph to suffer a severe blow. With the cheers and congratulations of the multitude ringing in the air around him, the projector of the astounding success sat bowed with deepest, heaviest grief at the bedside of his best loved son, who lay a-dying ! A sad stone to be put on the cairn of his memory ! The lights and shadows of our lives are but as the juxta-placed squares of the chess-board, after all !

THE STOCKTON TERMINUS.

CHAPTER II.

THE EXTENSION OF THE RAILWAY.—THE LIVERPOOL AND MANCHESTER LINE.—INCIDENTS OF THE OPENING.

THE Stockton and Darlington line was a marvellous success, though only at first used for merchandise and coal traffic. By degrees passengers came to the company, and were carried. As is well-known, this railway was the creator of Middlesboro'-on-Tees.

Notwithstanding the success of this first line, the stiff-necked Canal Company did not improve their ways between Liverpool and Manchester. So the line of railway was projected, and this one really marks the true commencement of railroad enterprise in this country. The means of its initiation were found in the same way as now is usual. A prospectus was issued in proper form, and a parliamentary committee sat on it for a considerable time.

The story of the Manchester and Liverpool Railway has been told so well, the evidence of Stephenson has been so often commented on, the details of the scheme have been so often described, that it is not necessary to go over the ground again very closely. But it is impossible not to mention the chief points in connection with the line and its historical open- ing, and the comments which the journals and reviews of the period made upon the project. The committee was sitting for the consideration of the bill while the Stockton and Dar- lington line was being constructed. The chairman came to the conclusion that a speed of three and a half to four and a half miles an hour could be done with safety. George Stephen- son had declared that twelve miles an hour could safely be attained !

The same limitation was made to the powers of the engineer. The crossing of Chat Moss was looked upon as the scheme of a madman. "Ignorance almost inconceivable!" "No engineer in his senses," said one C. E., "would go through Chat Moss if he wanted to make a railway from Liverpool to Manchester." We know how Stephenson triumphed. Thanks to Mr. Smiles, all young people are acquainted with the history of the conquest of Chat Moss, and of the opening of the line, after the celebrated competition of the locomotives.

At that period the types of locomotive varied very much. Trevithick had initiated the steam-horse; Hedley had im- proved it; and Stephenson had adapted it. Trevithick had unfortunately failed in obtaining recognition for his high- pressure engine. Blenkinsop ran his locomotive in 1812; and in 1813 William Hedley, who had already been engaged on the locomotive steam-engine, took out a patent for his smooth- wheeled engine—the true pioneer of our locomotives of to- day; and to Hedley, in our opinion, the real origin of the present-day railroad-engine is due. In "Puffing Billy" Stephenson adopted Blenkinsop's cylindrical boiler, and Hedley's idea of smooth wheels; but subsequent improvements were made by Dodds, and Stephenson, and Hackworth. Therefore by the year 1827–8 not only were locomotives different, but they were not held in any high esteem. Even on the Stockton and Darlington line the greater part of the haulage was done by horses, and it required all George

CHAT MOSS, OVER WHICH THE RAILWAY WAS CONSTRUCTED BY GEORGE STEPHENSON.

Stephenson's powers of persuasion to induce his directors to adopt the locomotive on the Manchester and Liverpool Railway.

The opposition which had assailed the railroad was now transferred to the motive power suggested. The line which was to ruin landowners, to which the Duke of Cleveland had objected because it was intended to run through a fox-cover; the railway to which Eton objected because of the smoke, which would blacken all the country round, and enable the boys to play truant, was derided. The *Quarterly Review* had deemed all these schemes visionary and unworthy of notice. "The gross exaggerations of the powers of the locomotive, in plain English, the *steam carriage*, may delude for a time, but must end in the mortification of those concerned. We should as soon expect the people to suffer themselves to be fired off upon one of Congreve's ricochet rockets as trust themselves to the mercy of such a machine, going at such a rate—viz., twelve miles an hour!" So said the sapient *Quarterly*.

This was the spirit in which the civil engineers not too civilly met George Stephenson's suggestions to employ the "locomotion" engine; and the directors, most intelligent men, could not make up their minds in the face of so much opposing testimony. Terrible evils, deaths, and danger were predicted; but after a visit to Newcastle the directors came to the conclusion that horses would be of no use to their line; and *stationary* engines were suggested.

With this intention celebrated engineers were consulted. Messrs. Walker and Rastrick pronounced against the locomotive, and while they were considering this verdict suggestions and hints flowed in upon the projectors. The most ridiculous ideas were put forward, from the invention to reduce the friction so low that the train could be drawn by means of a silk thread attachment, to the application of power sufficient to "rend cables asunder!"

Steam, gas, columns of water, and of mercury, with compressed air, and the application of the air pump to create a vacuum; "machines working in a circle, without fire or steam-generating power, at one end of the process, and giving it out at the other; carriages which conveyed their own railway (a plan since found feasible); "wheels within wheels to multiply power," and many other wondrous suggestions, tickled the ears of the directorate.

The result was the decision of the majority of the directors to run locomotives, and the famous trial at Rainhill was the result. A premium of £500 was the inducement. The con-

ditions of competition were published, and on the appointed
day, the 6th October, 1829, the long-expected trial took place.
Each engine was to make two trips, or thirty-five miles, and
the manner in which the judging was to be done was as
follows :—

One judge was stationed at post No. 1, and the other at
post No. 2. The level piece of railroad extended for nearly
two miles; one mile and a half being permitted for the speed-
run, and one-eighth of a mile at each end was allowed to check

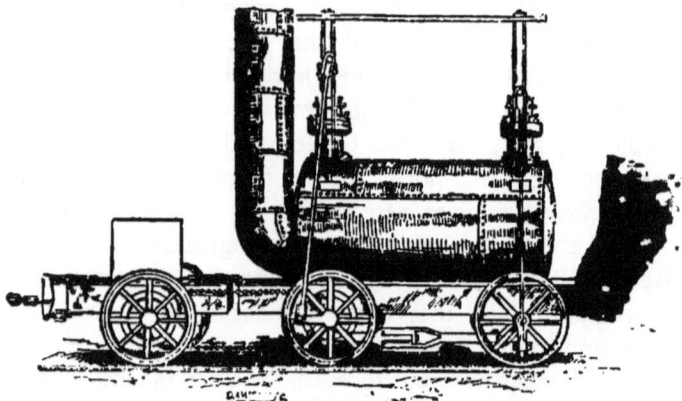

GEORGE STEPHENSON'S ENGINE, 1815.

the speed, or to get it up. The time was thus taken at full
speed on all the way allowed—one and a half miles.

Four engines competed on paper, but Mr. Burstall's *Perseverance* did not come to the post. It met with some accident
on the way, and the actual trial was carried out between the
Rocket of Stephenson, driven by one Charles Fox; the *Sanspareil* of Hackworth; and the *Novelty* of Braithwaite. The
result is well known. The *Rocket* bore off the prize, the
multitubular boiler adopted by Stephenson succeeded, and the
triumph of the locomotive engine was assured.

The scene on the occasion of the experiments was extraordinary. Every inch of space was occupied. From every

town and city came the expectant multitudes : the scientist, the
lounger, the pressman, the engineer. Some came to scoff,

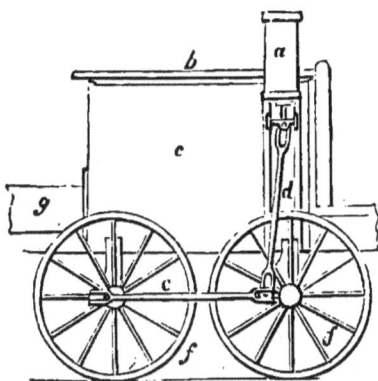

THE "SANSPAREIL."

others to applaud ; but both were of the same mind afterwards,
and agreed that the fiery horse was a factor in history and in

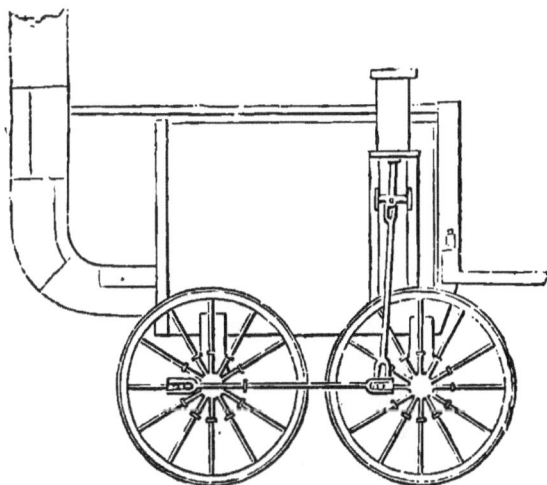

THE " NOVELTY."

commerce. For several days the "trial" trips continued, and
then came the final test, and with it triumph !

o

Just above twenty-nine miles an hour was the speed attained by the victorious engine.   This success sent up railway shares with a run almost as rapid financially as was the *Rocket's* in an engineering sense.   Fox, the engine-driver, afterwards came into notice in 1851, when the driver of the *Rocket* appeared in London as the equally famous builder of the Great Exhibition Palace, now at Sydenham—Sir Charles Fox.

Fox is not mentioned in many accounts of the Rainhill competition.   He had succeeded in life by his energy, though his early and rash marriage had caused him to be turned out of doors by the irate doctor, his father.

Amongst the names of the committee of the railroad, so soon to be opened, we find those of Robert Gladstone, Joseph Sandars, the Booths—to whom so much of the success was due—Hornby, Benson, Ewart, Sharpe, Hodgson, and Bisley; notable names, earnest of success.   The opening of the line took place on the 15th September, 1830, and never have we beheld such a pageant of its kind.   Many and various descriptions of it are given by writers accustomed to railroads and locomotives, not from the standpoint of the eye-witness, and therefore every account not in the newspapers of the time, or in private correspondence, must fail to convey to the full the intense excitement, the dramatic interest, and the tremendous impression created by the event.

The reporter has given us a clear sketch of the appearance of Liverpool, which was never so full of strangers from all parts of the three kingdoms.   "Never was such an assemblage of rank, wealth, beauty, and fashion !"   The area in which the railway carriages were placed "was gradually filling with gay groups eagerly searching for their respective places as indicated on the tickets."   After a while the railway cars were brought out ready to be attached to the eight locomotive engines ; flags waved, bands played ; the trains were distinguished by different flags, and a Grecian Car was occupied by the "Wellington Harmonic Band."

The hero of Waterloo arrived shortly before ten, when "the

discharge of a gun, and the cheers of the assembly," welcomed him. The band played the "Conquering Hero." The cars were then permitted to proceed gently down the hill to the Edgehill tunnel, where the engines were waiting. The tunnel was lighted with gas. On arriving at the engine-station the Duke's car was attached to the *Northumbrian* engine on the southern line, while the other cars were attached to their engines on the northern line.

All went well until the train reached Parkside, where the majority of trains passed the Duke's car. Many people alighted, although warned on their tickets to keep in the trains. There were other trains approaching. The *Rocket* was coming! Instantly a scramble took place from the line. Some ran off, some clambered into the carriages. Mr. Huskisson was one of the latter. He had mounted, when the door swung back. He fell on the line, and was mortally injured.

His leg was terribly shattered. The Earl of Wilton, Mr. Holmes, and others raised him up, and placed him in a carriage. He asked for his wife. "I have met my death," he cried. "God, forgive me!" He then fainted. Robert Stephenson drove him on his engine to Eccles, near Manchester. An immense concourse was awaiting the arrival of the procession. Eventually the trains were despatched thither, but all festivity was avoided.

"The *Phœnix*, with its train, was then attached to the *North Star* and its train," on separate lines, and from the two united a long chain was attached to his Grace's car (the Duke's, now engineless, for the *Northumbrian* was absent), and though it was on the other line of rail it came along very well.

But the reception of the injured man, who had been carried to Eccles Vicarage, is associated with some very curious, and perfectly true, circumstances which are not often related with reference to the celebration of the day. We have seen how the neighbouring houses were filled with guests anxious to witness the ceremony of the opening of the railroad.

In one of these houses, near Liverpool, was a party of

GEORGE STEPHENSON'S ENGINE THE "ROCKET."

guests, amongst them being a Mr. and Mrs. Blackburne, the

brother and sister-in-law of the host, a Member of Parliament. To the surprise of the company, Mrs. Blackburne, on the afternoon before the opening of the railroad, felt very uneasy, and was greatly affected by a presentiment of impending evil. She declared that her presence was required at home at the Eccles Vicarage, and notwithstanding the persuasions and even the ridicule of her husband and her relatives, she determined to return home. She did so, and took the canal boat—as in those days was usual—to Eccles.

When she arrived she was met with some surprise, and heard with mingled feelings of gratitude and disappointment that there was nothing at all the matter—that her presence was not required; one of the children certainly was rather "out of sorts," but there was not the least cause for any anxiety.

Next day, however, she was consulted by a gentleman as to the provision of local special constables for the protection of the Duke of Wellington, who was then very unpopular with the people. Mrs. Blackburne without difficulty indicated the men who would be willing to act, and though quite satisfied that her return had so far been useful, the presentiment of danger or disaster had not been fulfilled.

When the brave lady had thus assisted to provide the guard for the Duke's train, she made her own plans for seeing it pass, and in the tent she had provided the governess with the children and Lord Wilton's daughters took up their position with the vicar's wife. The day is described as threatening and thundery; and the gloom of the weather augmented the feeling of disappointment and apprehension which was gradually overspreading the watchers, who, lining the road, turned continually in the direction of Liverpool, in expectation of beholding the trains.

But none came. At length a commotion became visible at a little distance. An engine had come, and there had been an accident. The wounded man was being carried up to the vicarage !

As soon as Mrs. Blackburne had recovered from the terrible

shock and fear that the injured one was her husband, she hurried away, and perceived thankfully that the wounded man was not the vicar.　Mr. Huskisson—for he had been brought thither at the desire of Lord Wilton, and with his own consent —was immediately tended most carefully by Mrs. Blackburne. Her skill and her private medicine-chest were at once devoted to him, and though she could not, nor could the doctors, save his life, she soothed his dying hours and alleviated his pain. But for her presence and personal attendance the agony endured by the unfortunate Mr. Huskisson would have been so much greater, and he would not have been enabled to make the arrangements and to partake of the religious rites which Mrs. Blackburne's care and attention enabled him to do.

Thus her presentiment was fulfilled; not in the sense she had feared, for her own family, but none the less usefully. There is another circumstance connected with this most unfortunate incident which may interest those who see coincidences in these cases.　Mr. Huskisson, before he started on his first, and, as it proved to be, his last, railway journey, expressed a fervent wish that he were safely returned.　" I wish I were safe back," he said to his friend; and in that aspiration a fear of coming danger may be traced.

STEAM CARRIAGE.

ST. JOHN'S WELL, STOCKTON.
(The first rail was laid where the figure stands.)

CHAPTER III.

RAILWAYS AND THEIR EFFECTS.—THE SOCIAL ASPECT OF TRAVELLING.

T is not desirable that we should give any further particular account of the several railways which so soon came into existence after the successful open ing of the Manchester and Liverpool line. The names of the engineers who drove the engines have been pre-served.[1] Amongst them were George and Robert Stephenson,

---

[1] G. Stephenson drove the *Northumbrian*; Swanwick, the *Arrow*; Locke, the *Rocket*; Alcard, the *Comet*; Robert Stephenson, the *Phœnix*; and Gooch, the *Dart*.

Joseph Locke, Alcard, and Thomas Gooch, names afterwards associated with our great railroad enterprises ; the North Western, South Western, and Great Western Companies still preserve the memory of Stephenson, Locke, and Gooch.

The construction of the London and Birmingham railroad was quickly followed by others : the Great Western, the Grand Junction, the Essex line, and many more were rapidly proceeded with ; but even with the ever increasing facilities afforded by the Liverpool and Manchester Railway, the opposition to the London and Birmingham and other lines was strong and influential. The surveys had to be made at times when the owners of property were absent, and when parsons were at church. Northampton would not have the proposed line at any price, and so the contractor had to run the track through the hill in which the celebrated Kilsby tunnel remains a monument of the ability of Robert Stephenson.

The "men of influence and education," of whom Sir F. Head speaks, gained their

ROBERT STEPHENSON.

point, and turned the railroad aside. The tunnel, of which more in future pages, cost £300,000, and killed the contractor ! The influx of water made the tunnel a work of immense difficulty. Again, both Oxford and Eton distinguished themselves in opposing Brunel and his Great Western ; even Slough was prevented the enjoyment of a station on the G.W.R., which originally was planned to Reading only. Many worthy individuals hated the idea of railroads : Colonel Sibthorpe never would travel on one ; people have been known to make wills, and leave their relatives money only on condition that they would not travel by rail ; and in one case a testator carried his animosity so far as to will that none of the promoters of a certain railway, which he named, should ever be

ISAMBARD K. BRUNEL.

entertained in his house after his decease, nor should their sons have any food or drink there !

Notwithstanding so much opposition, railroads made their way. Even in London at the present day we may perceive the results of this opposition. Our London termini were all in distant districts, and some are still inconveniently situated. The Birmingham Railway engines were stopped in Camden

Town, and then only with ropes were the trains run into Euston Square. The same winding-up plan was, within the writer's recollection, in the tunnel entering Lime Street, Liverpool. The London and Southampton line was only completed to Vauxhall; the Great Western stopped at Paddington; the Northern and Midland at King's Cross and St. Pancras; the Kent and Sussex lines at London Bridge in the Borough; and so on. People feared the steam-engine. The North Western of the time and the Great Western were intended to unite at Willesden, but when Brunel introduced his seven-feet gauge, the severance occurred; and the broad gauge went on its own way into Paddington, rejoicing !

The great battle of the gauges was fought, and yet Brunel retained his position. The Stephensons deemed such a width "humbug" as great as the atmospheric line which we used to travel on from Kingstown to Dalkey, in Ireland, the air in the tube being exhausted by an immense pumping engine in a great house whose windows shook and clattered violently, and into which engine-house it was our delight and fear to be admitted when the huge machine was set in motion.

Within a few years the railroads, which, it will be observed, follow the old paths and roads pretty closely, began to turn off the coaches. The mail-contractors began to feel in a quandary; the Government was nervous; the railways took the passengers, the coaches had the contracts for mails, and when so many of the old "stages" were removed the letters could not be carried unless heavy payments were made. A select Committee decided to use the railways. The companies denied the right of the Crown to utilise private enterprise. Mr. Labouchere declared that the railways bound the land in bonds of iron, but eventually an arrangement was come to, and a subsidy was granted; the terms being settled between the interested parties.

Until 1836 comparatively few lines had been initiated, but then the railway asserted itself, and a kind of mania for such investments and speculation arose. Of lines to Brighton five were suggested, and the shares of existing railways bounded madly. Even in Parliament, amongst whose members railroads were not first favourites, attention was directed to the growing craze, although Colonel Sibthorp declared them "frauds" and "robberies," and used strong language concerning them.

GEORGE HUDSON, THE RAILWAY KING.

The social effects of this passion for the railway were curious. As the coaches died off, the industries immediately connected with them suffered. The inns were closed or, as related of one at Hounslow, took to selling "eggs and new milk." "New milk and cream sold here" was one legend quoted. The makers of stable brooms and brushes also felt the change. Clergymen complained that they could not persuade their parishioners in country places to come to church, because they preferred to wait and watch the trains.

"Canals dropped," and even the every-day conversation became interlarded with such phrases as "going like steam," "getting up steam," "run off the rails," and so on. But, of course, transit improved, produce was rapidly conveyed, and the supply of milk has since so increased as to become a hindrance to traffic and a nuisance to the passenger.

But, as may be believed, the railways did not have it all their own way. People made tremendous demands for land, and opposed the project in every way if not satisfied; and by degrees the reaction came, and railway shares fell. Then a commercial crisis occurred, and the usual miseries followed. With these events, Mr. Hudson first appeared—the Railway King.

It is on record in the newspapers that the Queen never travelled on a railroad until 1842! On the Waterloo Day, 18th of June, in that year, the Queen took her first trip on the Great Western Railway. This is a remarkable occurrence, because Prince Albert had at times used the line. Since that time—now fifty years ago—Her Majesty's journeys have been long and frequent; and the great Duke of Wellington, the "Iron Duke," who had held out so long, at length recognised the "Iron Horse" in 1843.

The terrible accident which occurred about this time on the Versailles railroad had an influence on British minds and lines. The day was Sunday, and the King's fête was to be celebrated on that afternoon and evening. When the display of fountains had ceased, the visitors hurried to the railway; and the train being unusually heavy, two engines were employed. One of these—the "pilot"—broke down, and was immediately run over by the other. The fire-boxes of both locomotives fell, and as the train could not immediately be stopped, several "cars," or coaches, caught fire. The miserable passengers had been locked in the carriages, and many were quite unable to escape. The scene, as described in the

Annual Register, was terrible, the only parallel to it being the Abergele accident in later years. More than fifty dead bodies were discovered, burnt almost beyond recognition. An English engineer of the train died from suffocation in his efforts to assist the passengers. The other engine-men were crushed to death

This accident led to a discussion in the British Parliament, and some lively times were enjoyed. One member informed another that he was not an impartial witness, inasmuch as if he were in a railway accident, and the carriage door was locked, he was too fat to escape by the window! Sir R. Peel did not think that legislation was necessary on the point; but another member attributed the accident to Providence, in revenge for the Sunday travelling, and he proposed that "no railway should be used on the Lord's day."

Such an agitation as might have been anticipated arose. Railway excursions on Sundays were termed, "trips to hell at seven shillings and sixpence"—a choice phrase which has had its parallel in modern days on the South Coast line, if rumour be true. "Solemn warnings for Sabbath breakers" were handed to passengers, and warnings of "God coming in judgment" met the weary ones who longed for fresh air and a glimpse of Nature's gifts beyond the smoky towns. . . . The amendment was rejected in Parliament.

But accidents were not very frequent, nor so fatal as the above. One amusing anecdote at any rate may be told of an occasion when a train in which an elderly lady was travelling ran off the line down an embankment.

A fellow traveller, a gentleman, looked for her anxiously when he recovered his senses. She had regained her feet, and as soon as possible asked,—

"Can you tell me if this is Salem?"

"No, madam," replied the man; "this is a Catastrophe."

"Oh, indeed! Then I hadn't orter to have got off here," she replied.

So much for the Americans'·side of the question as regards

accident; but their view of the effect of the railway is interesting, and worth quoting :—

"Twenty miles an hour! why, you will not be able to keep an apprentice boy at work! Every Saturday evening he must make a trip to Ohio to spend the Sabbath with his sweetheart. Grave plodding citizens will be flying about like comets; all local attachments will be at an end. . . . Veracious people will turn into the most immeasurable liars; all their conceptions will be exaggerated by their magnificent notions of distance."

This production has scarcely been verified, but undoubtedly the train and the locomotive engine influenced the people, for we find it described by a woman as "a long black thing, spitting out smoke and crawling along the ground; and when it caught sight of her watching it, it uttered a loud yell, and rushed into a hole in the hill!"

*　　*　　*　　*　　*

Perhaps a few dates will be useful in closing this chapter, giving the years in which the principal railroads were opened, after the success of the Liverpool and Manchester : —

The Liverpool and Birmingham opened in　.　1837.
The London and Birmingham opened in　　.　1838.
The London and Southampton opened in　.　1840.
The London and Bristol opened in　　.　1841.
The London and Colchester opened in　　.　1843.
The London and Dover opened in　　.　1844.
The London to Peterborough opened in　.　1850.
The London, Chatham and Dover opened in　1860.
The Midland to London opened in　　.　1868.

Concerning each of these railways, volumes have been written in their names of "Great" Western, Eastern, Northern, and so on. If we were to relate the history of the railroads of the United Kingdom and their romances separately, a dozen large volumes would be required. We will accordingly content ourselves with the general view, and look at the railway travelling in olden times before we touch upon the railroad mania and the Railway King.

NAVVIES CAMPING IN A WAITING-ROOM.

CHAPTER IV.

THE RAILWAY MANIA.—MAKING OF LINES.—THE "NAVVY."— SPECULATION, RUIN, REPENTANCE

THE continual development of the railway system in England and abroad led to one of the most remarkable panics in history. The South Sea Bubble was perhaps as disastrous, but we do not think that its effects were felt for such a period as were those of the Railway mania.

In the early forties people, encouraged by the dividends paid by the leading railway companies, regarded them as safe as Government securities. In 1842 money was easy, and this plentitude of capital, which only fetched a low rate to investors, induced the most prudent people to look round for new openings. Consols were over par ; and as the openings for railroads increased, public attention was directed to them.

In 1843-44 some fifty new schemes were hatched, and in the latter year (1844) so many new plans were pushed forward that Government was compelled to interfere, and bring in the Bill for their proper regulation as "Joint-Stock" Companies.

By this time the public and the press had recognised the value of the railroad, but they went to the other extreme. The locomotive engine, far from being the devastating and devouring monster which it had been formerly represented, was now a generous and good-natured giant, which would hurt no one. The dangers of the slips of iron, on which trains ran, and from which they would inevitably rush off, according to former opinion, were now put aside—the railroad was the safest, the best, the speediest, the most physically and financially secure invention in the world!

Railways were then the "triumphs of a period of peace"; the "emblems of internal confidence and prosperity." Railways were the universal panacea for all ills of the body politic. Wondrous employers of labour, great levellers, the true means by which different sorts and conditions of men could be brought into touch one with the other. "The artisan need no longer remain buried in the country, the agriculturist may find employment in distant places," and, concludes a writer of the time, "by railways the whole country may be, and will be, under the blessing of Divine providence, cultivated as a garden."

This is pretty well for those who had not so very long before been denouncing the railroad as the spoiler of the land, the scorcher, the fiery monster, which would devastate the kingdom, and demoralize the inhabitants.

Urged on by such articles as these, the public of 1845 began to bestir themselves. In January of that year sixteen new lines were registered, and April witnessed fifty more companies, all of whose shares were quickly quoted at a premium. Every one, without distinction of age and sex, put a finger into the speculation pie. Infants figured for thousands in the lists; servants and clerks vied with their employers in becoming holders of scrip. The sportsman and the parson, the idler and the churchman, the dainty lady and the cook, each and all went in for shares in the spoil.

The impetus given to newspaper enterprise at this period was surprising. There were, in 1845, only three newspapers devoted exclusively to the railway interest, but during the early part of the year no less than twenty were started, and others made ample fortunes out of the rain of advertisements which poured into their offices and into those of the ordinary newspapers. There were Railway *Worlds*, and *Globes*, *Engines*, *Telegraphs*, *Advocates*, *Examiners*, *Expresses*, *Reviews*, and *Standards*, which arose, lived on the crop of advertisements for a while, and then died in the cold shade of bankruptcy, in the winter of public discontent.

The advertisements were carried to all the papers in such quantities that extra sheets multiplied could not contain them, and ten thousand pounds sterling per week was a by no means extravagant sum to receive in payment at that period. Then other technical business flourished : the engineer, the draughtsman, the surveyor, the clerk, the outsider even, with the smallest smattering of engineering knowledge, came to the front, and made five and ten guineas a day.

Those were stirring times, and hardly any scheme was too wild for acceptance. It did not matter whither the line was projected the shares went up. Barren and blasted heaths, inaccessible mountains, almost impassable estuaries, and even places totally devoid of inhabitants had each their line. The " Glenmutchkin Railway " is not a greatly exaggerated description of the condition of affairs.

It seems almost incredible—did we not know by experience how easily the public runs away with a bit of speculation in its teeth—that sensible business men, warned by the *Times*, and other " cool-headed " journals, of the consequences, should have run such risks. Engineers ruled the world : they promised everything, crossed rivers, tunnelled mountains, and bridged chasms with magical readiness, on paper.

P

As to employment, it was at a premium. The navigator was triumphant. This peculiar specimen of the Railway natural history had in times past not given unmitigated satisfaction, nor proved an altogether desirable addition to the neighbourhood in which he had been let loose. His manners and customs did not entitle him to be the entertained of the villagers The "navvy" was and is still, to a certain extent, a member of a peculiar class of workmen remarkable for independence, reckless courage, hard work, and many like virtues, which are too often concealed beneath a rough, surly, cruel, demeanour ; and in the early days of railroads the representatives of the class were anything but desirable acquaintances. Introduced to history by the Duke of Bridgewater, the navigator increased and multiplied, and found his level in more senses than one ; unfortunately, rather a low moral level, which prevented him from enjoying many advantages.

These workmen lived together—"herding," says a writer, "like beasts of the field, owning no moral law, and feeling no social tie, they increased with an increased demand. They lived only for the present, they cared not for the past, they were indifferent to the future. They were heathens in the midst of a Christian people, savages in the midst of civilization."

A perfect terror was manifested in many parts of the country where these men were at work, and no wonder. Perfectly devoid of all education, and brutalized to a degree we now would regard as impossible, the navigators acted upon their own impulses when not actually compelled to be at work. Under the circumstances is it wonderful that game disappeared, that hen-roosts were rifled, that even more serious crimes were committed, and that the perpetrators escaped by presenting a solid and unflinching front to any posse of police that dared to invade the horrible sink of debauchery and criminality they termed a "settlement." The way in which the farmers were treated, the manner in which their children often suffered, were sufficient arguments against the presence of the navvy, and no wonder that railroads were regarded askance by the quiet country people, when the flood of speculation was let loose in the '45.

But the prophecies of the steady ones caused no check. Money became wildly sought after. No one was satisfied with anything under ten per cent., and so business was neglected and every one rushed into the market. One of the greatest

SERVING OUT OATMEAL AND WATER TO NAVVIES.

of Railway Kings uplifted his voice, and Mr. Hudson, whose marvellous career is still a by-word, warned the investors that degeneration would surely supervene when competition set in. But they heeded not the writing on the wall.

The extraordinary impudence of the people entrusted with the surveys are commented upon by contemporary writers. Londoners were kept in a constant state of anxiety by rumours as to the course of the projected lines. "Young gentlemen with theodolites and chains marched about the fields ; long white sticks with bits of paper attached were carried ruthlessly through fields, gardens, and even houses." In vain country people protested. If they didn't like it they must get used to it, was the polite rejoinder, and one remonstrating, was informed that if he didn't " like the railway through his land he would have it through his kitchen ! " Such were the amenities of the mania.

Throughout England the madness had spread, and many large towns were almost as bad as London, where the three Stock Exchanges were crowded, and where all the efforts of the police were needed to keep the streets clear. The racing and chasing, the fearful worry and anxiety were all visible at once, but triumph was dashed to the ground shortly after, when it had risen higher than usual.

So the fearful excitement grew. The Board of Trade had fixed the 30th November as the date on which all plans must be deposited, and the fearful racing and rushing to complete it are impossible to be described. Already signs of a decline had come in October. Hundreds of men, who had literally nothing, adventured thousands, and rendered themselves liable for payments which they had neither the means nor the intention to make. Some of these speculators made large sums by selling their scrip, and the return demanded by the House of Commons shows the variety of people who adventured. The duke and the deacon, the warden and the widow, the soldier and the sailor, the surgeon and the solicitor, the dean and the doctor, the priest and the puritan, all came in, and strove to struggle out, with Mammon for their friend, with their broker and banker.

The Bank of England raised the rate of interest on Thursday, 16th October. This had an immediate effect on the markets, for money became tighter and the war could not be carried on. Would people pay ? Could they obtain the money to close their bargains if the rate advanced ? were important questions. Consols fell, and most other securities which take their tone from the " Simple Threes " declined. Then began a panic. What will happen if I cannot realize? was the cry. The shares which had been worth £100 yesterday may be almost waste paper to-morrow. The allottees in many cases thought it prudent to refuse the allotted shares, and to disappear.

Reaction had set in! The railroad, which had been the benefactor of the world, was now a poor thing—a very rogue, one of whose society or association the individual was heartily ashamed, and an acquaintance to be denied more often than Peter denied his Master. No one knew the railroad. It was a mistake to suppose that any respectable person had had any dealings in scrip; but all this denial could not avert consequences.

We may deny having been in questionable company, or in association with those who are sick or infected; but when the disease breaks out, and we die or are crippled for life, it is of little use our having entered protests. So in this Railway story. Hundreds were "crippled" indeed, many committed suicide, thousands were cast upon the world penniless, children were ruined, and parents imprisoned or starved. The unrelenting Juggernaut of shares came crashing along, and passing over the helpless victims slew them ruthlessly.

But the railways which could lodge their plans and proceed to construction might pull through and succeed eventually, and hence the anxiety to comply with the rules of the Board of Trade. By some oversight the last day had been named for Sunday, and consequently some difficulties arose. Lithographers and printers worked double tides, some worked all night and hardly rested for many days. Trains were bespoken. Many specials were ordered on Sunday morning early to bring up the plans for other lines. Post horses had been long engaged and were as rigidly guarded as Derby favourites to prevent a rival schemer obtaining possession of them. Again some of the railroads declined to convey the rival plans and the clerks in charge, so chaises, carts, waggons, any possible or impossible route was sought and examined, to give the needed access to the Metropolis in time.

That Sunday was a busy one. Mid-day was the hour named for the *closure*, and as it approached the excitement became intense. The clerks appointed to receive the plans were utterly incapable of doing so ; the rolls rained upon them. The hall was crowded with applicants, who were quieted when informed that each was in time. Then a post-chaise drove up. The clock was striking. Only just in time ! The door was shut, and those who had not arrived were lost !

But about ten minutes later another post-chaise arrived, the horses steaming. Three men leaped out each carrying a Titanic roll. They rushed down the entry to the door jostling and crushing each other. But the door is fast—"Too late ! ye cannot enter now ! " One man pulled the bell furiously, the inspector replied that the plans could not be received, where-upon they were thrown in at him, burying the door-keeper, who promptly threw them out. Again they were launched into the hall, and again dismissed ; and the door being then closed, the disappointed gentlemen, whose horses or whose postillion had failed them, had only to return crestfallen to their Surrey home.

The Government soon interfered on behalf of those com-panies which had been wrecked. A " Dissolution Act" was passed, and by bringing the affairs of companies before the public, unveiled their unsoundness; directors and promoters were proceeded against, settlements were made, and the subscribers were relieved of their liabilities, but at a terrible loss of money, which could not be recovered from the promoters and directors, who carefully hid themselves, or declined by any means to repay the subscriptions which they had wrongly ad-ministered or squandered. The Railway mania desolated many an English home ere it died out.

During this memorable mania many curious incidents took place. Disputes, and even pitched battles, between the land-owners and the surveyors were common. Day after day the intruders were beaten off, to return, as persistently as so many flies, to annoy the landlords. The appearance of the roads and railways as the day of lodgment approached will be long remembered.

Special trains, coaches and four, post chaises, all vied with each other. The Great Western sent up a special with such haste that the engine collapsed at Maidenhead and refused to proceed. While the driver and fireman were engaged in coaxing their refractory steed to continue its journey, another special with a swift flight "pitched into" the rival train, whether of set purpose or by the merest accident did not appear.

On the last day, the Sunday aforesaid, there were ten specials on the Great Western, and nearly twice as many on the present Great Eastern system—then the Eastern Counties. The means and shifts employed to pilfer or destroy or circum-vent the plans, designs and effects of rivals were legion. Theft, chicanery and bribery were all employed. On one occasion when a railroad had refused to convey the plans and clerks to London, the promoters of the new line, determined not to be beaten, got up a pretended funeral. The hearse and mourners with a coffin were all duly prepared, and carried to town, with all the outward signs of woe decorously instigated. The hearse and attendants reached the metropolis in safety, and were driven to the last resting-place of the coffin, which contained the plans for the rival line, and was deposited at the offices of the Board of Trade.

Until midnight the scene there almost baffles description. The efforts of the agents to deposit their plans in the office exceeded even those of the most persevering cuckoo to leave her egg in a strange nest. The crowd smashed the windows and hurled plans in. Of course men bearing the same surname abounded, Smiths and Joneses when called rushed in, con-tended for precedence of hearing, and almost fought for it. On one occasion, when the agents arrived just after midday, the delay had been caused by the post-boy who pretended not to know—or really did not know—the way to the office, and the agents who had come up from Essex at full speed were beaten "on the post" by a minute or two.

We read now-a-days of wondrous express speed in railway travelling. On the celebrated occasion, when England and the United States were trembling on the verge of war over the Slidell and Mason incident, the despatches were carried up

EARLY GREAT WESTERN ENGINE (1838) WITH DOMED FIRE-BOX.

by the London and North Western engine from Holyhead to Euston in five hours. The distance is 264 miles, which gives the average speed as 53 miles an hour. To effect this rapid delivery an engine had been kept constantly "in steam" at Holyhead from the 2nd to the 9th January, 1862. The Atlantic cable did not then exist, and this feat of bringing the despatches was, and is, still regarded as remarkable for such a run; but in the days of the railway mania there was a train with plans, etc., which ran up to London from a country town, 118 miles distant, in an hour and a half, or at the rate of 80 miles an hour.

We may conclude our references to the railway mania by recording the number of schemes brought forward in the '45. There were six hundred and twenty railways projected, which, if they had been carried to completion, would have involved an expenditure of at least five hundred and sixty millions sterling. This number does not include the abortive plans which never saw light, and which are recorded by the *Times*, in 1845, as amounting to no less than 643.

Two hundred and seventy-two plans were sanctioned by the Houses of Parliament, and the unfortunate shareholders, as already explained, had to find the capital or be ruined. Hundreds were ruined !

PRIMITIVE TRAIN.

CHAPTER V.

EARLY RAILROAD TRAVELLING.—THE "COACHES."—THE
SIGNALS.—ANECDOTES OF TRAVELLING.

IN considering the railroad travelling of the past, we must remember that the ideas of the early railroad-makers had not got beyond horse-tramroads. The Stockton and Darlington line, and even the Manchester and Liverpool railway, were at first intended for horse haulage only. The early locomotives were small. Stephenson's engine, "Locomotion," which inaugurated the Stockton and Darlington traffic in 1825, weighed only eight tons; and the engineer was satisfied if he could "knock sixteen miles an hour out of her," even at the risk of heating the smoke-box to a red heat; though the celebrated *Rocket* subsequently made a pace of fifty miles an hour.

From the successful opening day of the Liverpool and Manchester line, passenger trains began to run. The first "regular" train started on Friday, September 17th, 1830, and any one who was on the look-out for omens might have anticipated evil of the commencement of the traffic upon such an inauspicious day. The trains did not run frequently at first, because, though curiosity had influenced many, the realized danger of "playing" with such formidable monsters had an effect. From the opening of the line to the 31st December, 1830, the number of passengers carried was only 71,951, and during the whole year 1831, 445,047, which rose by degrees to 522,991 in 1836 ; and on an average it was calculated that each inhabitant of Liverpool, Manchester, and Warrington travelled once a year by the railroad.

It must be confessed that the accommodation in the early days of railroad travelling was not calculated to induce people to go by the trains. The mail trains, when introduced, were fairly comfortable ; but the general public were not expected to travel otherwise than by stage or waggon, so the first railroads and their charges were rather calculated to meet the requirements and the purses of the rich. The rough method of coupling the "coaches," the somewhat uneven track, the small wheel-bases of the carriages and engines, gave rise to much discomfort and oscillation. The jerking, bumping and shaking in starting, stopping and progressing were great, and trying to the nerves and frames of humanity, and of the "coaches."

The railway carriage is still in the service termed a "coach," and a glance at the form of it will assure the spectator of the origin of it. The highway-coach was the pattern, and it is still adhered to practically in England, save when Pullman cars, or certain elongated saloon or third-class carriages are employed. The "coach" stands confessed.

In those early days each "coach" had its name painted on it conspicuously. The coach was also of brilliant hue, and the writer can remember the time when the different classes of carriages were painted distinguishing colours. The first class of the first line with which he was acquainted was painted a deep royal blue or purple tint, the handles were brass, the seats padded and covered with blue cloth, the blinds were to match, and one felt one might, even then, travel in such cosy coaches until "all was blue." The second-class were of more sober green, or brilliant gamboge colour. Some were open carriages with the merest protection of seat rail, and the enterprising schoolboy might, and as a matter of fact did, seat himself upon the upper foot-board (for there were no doors windows, or any shelter save the light, low sides and roof), place his feet upon the under foot-board, and so contemplate the line and the sea! One lad, to our own certain knowledge, occasionally rode on the buffers at night to "find out how it felt," and on one memorable occasion when a slight check caused the buffers to approach each other more closely, a pinch of a severe character conveyed painfully the information ostensibly desired.

FIRST RAILWAY PASSENGER COACH, 1825.

The third class were of a blue tint, but did not in any way approach the splendours of the first-class. The seats were hard. The second-class carriages had no arms, but were cushioned, and had windows at each end from which the line could be surveyed at leisure. These were closed carriages, of course. The spring was inaugurated by the "open second," which were very airy, like those on the top of the Paris suburban railway, but on the same "trucks" as the closed carriages.

The windows at the ends of the closed second-class carriages enabled the youthful travellers to observe the process of engine-driving, for on the down journey the "seconds" were in front, and in the rear on the way up. The engineers were generally affable, and many a ride did the writer enjoy on the little "tank engines" on the Dublin and Kingstown Railway !

The naming of the carriages gradually disappeared ; perhaps the accident of placing a distinguished French general in a "coach" named "Waterloo"—which *contretemps* did actually occur on one of our English lines—may have contributed to the abolition of the nomenclature which still continues to obtain in the Pullman vehicles, and on the locomotives of many lines. The heroes of heathen mythology and the gods and goddesses have always been favourites ; we should like to know how many *Jupiters, Polyphemuses* and *Cyclops* there have been recorded in the baptismal registers of the running-sheds !

The early carriages were not comfortable ; luggage was carried on the roofs, and the writer can recollect seeing charred portmanteaux and burnt tarpaulins drenched with water to extinguish the flames communicated to the luggage from the engines. The porters in those days swarmed on the roof and hunted out the luggage of passengers, using the iron steps as ladders. At such a junction as Stafford, the confusion was frequently very great. The mail guards were glorified individuals—in scarlet coats, with black belts, at times—like their predecessors on the mail road coaches, sat up on the roof in all weathers, guarding the mails. Indeed, the difference between the mail road coach guard and the rail road coach guard was then not great. There were trains exclusively mail and first-class, in which seats were booked and numbered ; a kind of limited mail, in which one could reserve seats—four in each compartment—and these were capable of being turned into 'bed-carriages" at will.

The lighting of the trains was from the outside. Huge lamps, suspended like road carriage lamps, shed their gleams through these "glass-coach" windows which were used principally at night, when the railway companies had at length ventured to run night trains ; for traffic at first was suspended

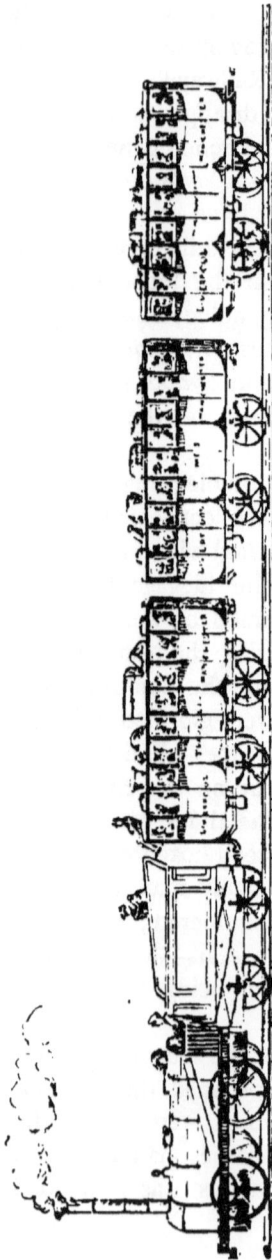

FIRST-CLASS TRAIN ON LIVERPOOL AND MANCHESTER RAILWAY, 1837.

during the hours of darkness, in consequence of the difficulty of signalling.

There were no regular signals at first. A "policeman" would stand with a flag; the red hue, most naturally indicating slaughter and death, was chosen as "danger" signal. As a compromise between the danger and the perfect white of safety, green (at times, blue) was adopted; and these colours have been continued in the latter-day lamps. In fogs or thick weather, drums were employed, on which the man perched by the line could beat a sonorous tattoo to warn the approaching train.

In tunnels the signalling was defective, and writers of the period could not grasp the methods by which this was to be effected. One scribe says:—

"It has been gravely talked of lighting tunnels artificially, so as to supersede the necessity for daylight. How or by what means this is to be done remains a secret. To philosophers and practical men, the hopelessness of approaching the solar by any artificial light is well known.

"Coarse as our optical means are in judging of degrees of light, it would be impossible to have a sudden transition from solar to lunar light without producing the sensation of great darkness. But the transition from light to darkness is not nearly so bad as the contrary, from the intense tenebrosity of a tunnel to the full, broad glare of daylight."

This transition must have been specially trying to the outside passengers, who, attired in the blue coats and white pantaloons of the period, mounted to the roofs and, furnished with wiregauze spectacles, endured the showers of grit, sparks and dust with praiseworthy patience; while a rain of ashes, which would have buried a village, fell upon them and off them as water from the feathers of the proverbial duck.

Nice travelling, you will say, was this, and inside the " coaches "— the late-introduced " seconds "— matters were not much better. There were no windows in those coaches. The vehicles were then only about eight and a half feet long, and four feet four in width.[1] The difficulty of squeezing in, and the discomfort, may be

[1] The actual dimensions were 8 ft. 7½ in. long, and 4 ft. 4½ in. in width. Seat 15 in. wide, door 18 in.

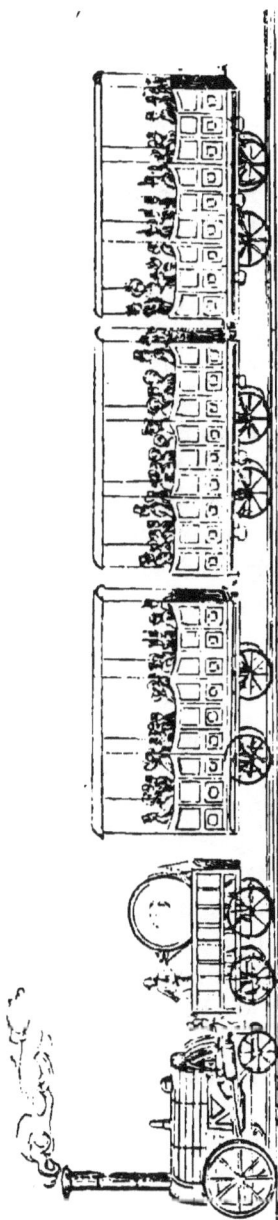

SECOND-CLASS TRAIN ON LIVERPOOL AND MANCHESTER RAILWAY, 1837.

imagined. The first express to Exeter took the first glazed second-class coach.

Again, in those days, passengers were not very respectfully treated. They were herded like cattle, locked in, and left to sit or stand, shaken about into each other's arms, jerked off their feet, and ordered about finely! Here is a notice which will illustrate the haughty demeanour of the Board of those days :—

"Passengers intending to join trains at any of the stopping places are desired to be in good time, as the train will leave the station as soon as ready, without reference to the time stated in the tables ; the main object being to perform the journey as expeditiously as possible."

There is a charming candour about this ! The "main object" one would have thought would have been to accommodate the public, by whose custom and favour the railroad people lived, moved, and had their being ! But not so. "The trains will start when ready !" yet, if no passengers were there—at first, it could not be ready to start, and would soon cease to run. The directors did not appear to mind that. Then come some warnings :—

"Passengers will be booked only conditionally upon there being room on the arrival of the trains, and they will have the preference of seats in the order in which they are booked. Each passenger's ticket is numbered to correspond with the seat taken. All persons are requested to enter and alight from the coaches invariably on the left side, as the only certain means of preserving accidents from trains passing in the opposite direction."

It will be evident to the reader of these rules that the number of passengers were very limited in those days. Fancy a Bank-holiday mob searching the compartments for the numbers corresponding to their tickets, or politely standing aside to permit the women or men, who had obtained tickets before the crush came, to take their places in the order of precedence in which the tickets had been issued ! But it is a question

whether we have not abandoned control on our railways too
much.   Some judicious regulation of the traffic on busy days
in our large stations is necessary.   Officials act as if nothing
unusual will occur on public holidays and race days.   *Laisser
aller* and *laisser faire* are all very well ; but it is at the *laisser
revenir* when the pinch and the push should be chiefly regulated.

The fares by the old mails were dearer than on other trains.
This relic of an old custom is still preserved by the North
Western Railroad, which books passengers by the Irish mail
trains only to certain stations at "express fares."   The first
compartments of leading carriages in these first-class trains
were reserved for male servants, and the second for women
servants, in attendance on their masters and mistresses, at lower
fares.   The mingling of the sexes was not apparently permitted
on the railways to servants.

The tables of fares were as follows, for instance :—

GREAT WESTERN TO MAIDENHEAD.
22 *Miles.*

		Fares.		Rate per mile.
1st Class—		*s. d.*		*d.*
Mail	6 6	. . .	3·54
Coach	5 6	. . .	3·
2nd Class .-			
Close	4 0	. . .	2·18
Open	3 6	. . .	1·90

Here we perceive the difference and the distinction between
the mail, the ordinary coach, and the closed and open carriages.   The third-class or "Stanhope" truck was simply a
cattle truck, in which no seats were provided.

In those, as in much later years, the railway companies did
not recognise the idea of the populace, the third-class travellers ; railroads were for the great.   Indeed, to put one's own
carriage upon a truck, pay second-class fare, and ride on the

Q

rail with your own coachman and footman on the box of your
road-coach — the family-coach — was quite an usual amusement, or diversion.   The ordinary travellers stood behind the
glowing engine in ordinary trucks—witness Turner's " Rain,
Steam, and Speed," in the National Gallery, wherein the red-
hot smoke-box is intended no doubt to accentuate the speed
at which the miscellaneous crowds of huddled-up passengers
are being whirled in open " boxes " through the rain, possibly
not far from Maidenhead Bridge.

The unequal fares had no relation to the cost of the line.
This seemed to people of the time a grievance.   " There is
no difference in first-class fares," says a writer, " between a
line which cost £42,000 a mile, and one which cost £10,000
a mile.   The directors," he declares, " have simply fixed the
highest rates which the public will consent to pay, or upon the
principle maintained in the post office, of determining what
quantity of traffic will yield as much profit as they require at
the rates which they choose to fix ; and neglecting or resisting
an equivalent increase at a lower rate, because they are unwilling to incur the risk of loss by providing more carriages
and more extensive accommodation."

Even in those early days it was demonstrated that the lower
the fares the greater the traffic.   " There can," says the *Popular
Encyclopædia*, " be no doubt that if the present high rate of
charge be maintained, the monopoly which railways possess
will prove a great obstacle to increased travelling."   Now, at
length, the British Railroad manager has risen to the recognition—as the Government did when they passed " penny post-
age "—of the patent fact, that the cheaper the means the more
popular the usage.   From year to year the improvement in
the third-class carriage has progressed, and now even in the
southern districts (S.E., S.W., and S.C.R.) it is more comfortable, in some trains, to travel " third " than " second."   The
latter are dear, dirty, and dilapidated in most cases ; the
former are cheap, often much cleaner and more spacious,
and are no more crowded than the second-class ; for " five a

side" is the rule in both cases. Some railway companies complain of the falling-off in second-class traffic ! Let such companies put only four passengers a side in the "seconds," and brush up the cushions, and they will find plenty of people ready and willing to travel "second." But while the middle-class people are so ill-treated in this, as in so many other ways, directors must not be surprised if they go for the third, in which they are considered.

Passengers, in the days when the railway was young, were sorted and locked into carriages for their destination—a mode of proceeding which called forth many bitter complaints in the newspapers, and gave the celebrated clerical wit the opportunity to say that the directors would never give up the pernicious habit "until they had caused the death of a Bishop — even Sodor and Man will do !" concluded Sydney Smith. . . .

*   *   *   *

But to return to the signals. The first idea of lamp signalling was a tallow candle, which a station-master put in the office window when he desired to stop a train. This quickly developed into the use of a lamp ; and after the opening of the Liverpool and Manchester line the ground-light became elevated, and was put upon a post. In time the post itself was raised ; and the semaphore by day, and the lamp by night, gave the necessary protection. The semaphore had been in use for other purposes in former days ; but, as railway enterprise extended, the system of signalling by "telegraph," as it was called, came into use on the line. In 1841 it was adopted on the railway, with levers on the post, to which an employé of the company attended, his hut or cabin being close by, for the now universal wire and the electric current had not been thought of.

Necessity is the mother of invention, and laziness, or a wish to save oneself trouble, has ere now resulted in improvements. As the desire for play led Master Humphrey Potter to devise the addition to the steam engine, which resulted in the effective condensation of the steam by the machine itself, so the wish to save his legs induced a railway signalman in Scotland to fix a wire and pulley with a rough counterpoise to one of the two signals intrusted to his care.

By these very simple means he was enabled to work the more distant signal without leaving his cabin. From this the present method has been gradually evolved, and the manufacture of railway signals, with their elaborate "locking" apparatus, by which the points and the signals work simultaneously, is an art. The invention of the "switch" on the line is due to Sir C. Fox, the same engineer who has been already mentioned as the driver of the engine *Novelty*, in the memorable trial trip at Rainhill. The semaphore was introduced in 1841, the inter-locking came in a few years later; but the system, as at present adopted, did not take root until Mr. Saxby fitted it on the South Eastern in 1856. To thoroughly understand the working of the signals the enquirer should obtain permission to enter one of the "cabins" near a large and busy junction, and he will then be so confused at first that he will think railway signalling work is on the high road to the madhouse. The banging of the numbered levers, or the ringing of the bells is to him only noise—sound and fury signifying nothing!

But let him wait and watch. He will perceive, after a while,

JUNCTION SIGNALS.

that order is regularly evolved out of seeming chaos. The levers with certain numbers on them are moved, and the levers indicated by those numbers are also moved, and no others, for that train. These levers set the required points and signals, and no others. The points cannot be moved without those particular signals, nor the signals without the corresponding points. Thus safety is assured. The novice may make a mistake, and open the wrong points, and lower the wrong signals

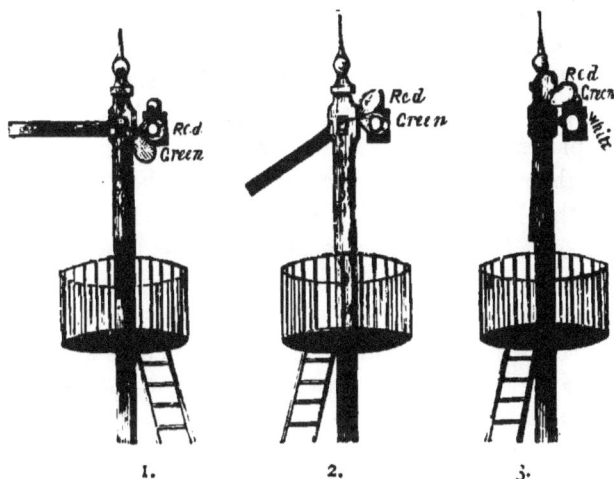

POSITIONS OF SIGNALS IN WORKING.

1. DANGER.   2. PROCEED.   3. BACK VIEW OF SIGNAL.

for a train, but no harm will result because the engine-driver will, if he be careful, pay no attention to any signals but his own.

In every signal-cabin are miniature signals which repeat automatically the movements of the distant " arms." Small discs inform the observer whether the signals be " on " or "off." Some of these discs and miniature signals are electrically influenced and worked by men in the *next block cabin,* and show how their signals are ahead. The men in A box cannot alter

them ; the man in B box holds the power over A's danger-sig-
nal ; and until B is satisfied, and has seen the train pass him,
he will not let A send another train on over the intervening
distance or " block."

The visitor to a railway signal-box will perceive that the
levers are painted different colours, and this difference of hue
has an important signification. In the boxes we have visited
by favour of the railway authorities, the up, down, home,
and distant signals have different coloured handles or levers,
so after a little practice and observation there is no chance of
making a mistake.

Again, the different numbers have each a signification. On
one lever, let us call it No. 20, you will find three other numbers on plates, very plainly and clearly cut. A very little search
will find them, with their corresponding numbers. Before the
head of the little family, "No. 20," can do his work, his off-
shoots, marked in plain figures, must be pulled over; and thus,
the points being set, the No. 20 can be pulled and the signal-
arm moved.

Thus you will observe that no train can come along before
the points are ready for its passage, and now no accident can
occur by the upsetting of the train as it is passing the points
and bars. On some occasions the signalman has caused an
accident by pulling over the lever before all the wheels had
passed ; but the latest improvements include the use of a bar
which it is necessary to pull up before the points can be moved.
As the flanges of the wheels effectually prevent any such upward
movement, the points cannot be shifted while any wheels are
on the rail points. This interlocking bar is a perfect safeguard.

It may not be generally known that the common red handkerchief, so often seen loosely tied round the necks of railway
porters, platelayers and signalmen, has been used as a danger-
signal on many occasions, and proved useful. We have read
somewhere that the neck-kerchief is the origin of the danger-
signal. Several anecdotes are told of the uses made of this
gear in placing it over a lamp to obviate collision, etc. But

even this serious subject has its comic side. Mr. Williams relates the following :—

"An Irishman walking along the Great Western railway was perceived waving a red handkerchief very vigorously, and as this was rightly regarded as a signal to stop, the engineer pulled up as quickly as possible, and the guard as well as passengers enquired anxiously, 'What is the matter?' Many people were greatly alarmed, and Pat was closely questioned.

"'There is nothing the matter, sir,' replied the Irishman, touching his cap ; 'but would your honour give me a bit of a ride!'

"This was adding insult to injury ; but the guard acquiesced and took up Pat into his van, greatly to the delight of the passenger, who congratulated himself upon his ruse. But his self-complacency did not last long. On arrival at a station the guard quietly handed Paddy over to the police, and no doubt he received some punishment adequate to his offence."

Another anecdote respecting the use of the platelayer's hand-

OLD SIGNAL POST.

kerchief has never been related, as far as we are aware, in print.

A platelayer was one warm day overcome, by the heat perhaps, and was returning home along the line, closely followed by a faithful fox terrier. The master stumbling along fell across the rail, and all the efforts of the "faithful hound" could not move him. In vain he barked, no one came, and at length the quadruped lay down beside the biped, and watched him.

Presently a train was perceived by the dog to be approaching upon the line across which the platelayer was extended—senseless, unconscious. The poor dog recognised the danger, and vainly tried to pull his master from the track. All his efforts in that direction failed. The train was approaching, fortunately not rapidly.   Suddenly an idea seized the animal.   Pulling his master's red handkerchief from his gaping pocket, the fox-terrier ran in the direction of the advancing train, fluttering the impromptu danger signal.   The engine-driver perceived the ruddy signal, pulled up, and the life of the platelayer was saved.

N.B.—We do not vouch for this.

Life in a signal-box, in a quiet country cutting, or on a breezy embankment, may be pleasant enough to a contemplative mind. If not exactly far from the busy hum of men and engines, it is not distracting.   But when the signalman is transferred to the bridge of a terminal station in which are more than two hundred levers, with perhaps a dozen men and lads to work the incessant traffic, the romance of signalling is apt to disappear !

THE POINTS.

Many curious tales are told of the sharpness of signalmen, and occasionally of their sleepiness.   The signalman on the Chester and Holyhead line once noticed the men on a passing engine were both fast asleep.   He telegraphed on, the engine was shunted as gently as possible, and the men rudely awakened to the damage; but the course pursued averted a catastrophe.   Another man has been known to go fast asleep in his cabin, and remain so in defiance of all the engine-driver's whistling.   So the engineer descended, entered the box, found · the man asleep, and departed, with his clock, on the way to his destination to report the case.   .   .   .

The station and cabin clocks are regulated daily by Greenwich time on all large railways, though there was once a "puzzle" in the pages of *Punch* " to find the time at Waterloo Station," a somewhat difficult operation at times; for the clocks, if regulated, seldom agree, and are often in flagrant disagreement with others on the line.

We will now look back again to the early days of travelling, which we have somewhat neglected in our brief notice of signals and signalling.

INTERIOR OF SIGNAL CABIN.

PRIMITIVE TRAIN EMERGING FROM TUNNEL. (LIVERPOOL AND
MANCHESTER RAILWAY.)

CHAPTER VI.

TICKETS, THEIR MANUFACTURE AND USE.—TICKET
SWINDLES.—THE RAILWAY CLEARING HOUSE.

THE first requisite for a journey by railway in the
early days was, as now, a ticket, then a mere slip
of coloured paper torn from a book, and stamped
with a round ink-stamp with the date, as a letter
is. The writer once had quite a collection of these flimsy
passes, but they have disappeared into the unknown. Often a
regular "way bill" was made out from the counterfoils of the
"tickets" issued, and handed to the guard, who thus became
acquainted with the number and the distinctions of the occupants of the carriages. The trail of the customs of coaches
was apparent for a long time on the railroad track.

Some people preferred to have their own carriages mounted
on the trucks and secured, and travel thus exclusively at second-
class fares. Such a sight is seldom or never witnessed now-a-
days, as the expense connected with such proceeding renders

it a luxury, the enjoyment of which is more than counter-balanced by the shaking and the cost.

The railway ticket business is considerable. The printed ticket was invented by a clerk on the Newcastle and Carlisle line, named Edmondson, in 1837. But long after that paper tickets were in use still; and we believe that not until 1850–51 was the present form of railway ticket adopted. The printing of tickets is a very important business on a big line, as millions are made in a year, and the old tickets are used largely in the process.

The cardboard of the required thickness having been provided, it is coloured in a machine which passes the sheets underneath coloured flannel. These coloured sheets of pasteboard are quickly cut and separated into tickets, as yet unprinted, of course. The method of printing is rapid and ingenious. The "blank" being sent over a space which the movement of the machinery fills up by the type. As the tickets pass on in a never-ending procession in single file, the stamp prints and numbers them consecutively, in a manner similar to the printing of the bank notes.

One very ingenious trait of the machine is related by a writer in *Cassell's Saturday Journal.* The tickets are sent along in sequence, and all pass muster if of proper size. "But," says the writer referred to, "tear off a piece from one of the blanks, put it in with the rest, and see what happens. It goes jigging away down the spout with the others, and the machine is pegging away as though it were far too busy to stop for anything, and you expect to see the maimed ticket pass muster with the others. Nothing of the kind however occurs; as soon as the spoiled one is pushed out, the machine stops dead, and it positively refuses to go any further until somebody comes to put things right."

The millions of tickets are counted and forwarded as required to the different stations; all sorts—single, return, excursion, etc.—of all classes, and of so many colours that Joseph's coat would pale with jealousy if it could see them. These

vouchers are two and a quarter inches long, and an inch and
a quarter wide.   The first ticket machines were made by
Thomas Edmondson and a friend, Blaycock, a watch-maker,
and used on the Bromsgrove line in 1840.

The mode of issuing tickets is familiar to all, and many
people regard them as an unnecessary wrong, particularly when
the examiners are exacting.   But great care is absolutely
necessary in the issue and collection of tickets, which have to
pass the clearing house if used over another line, as "through
tickets."   Besides, if checks and inspections were not carefully
organized and unexpectedly made, such is the tendency of
humanity to "err," that some travellers might be found in
compartments, for the enjoyment of which they had not paid :
or even without any ticket at all.

Sad as it is to contemplate, the proverb to the effect that
"to err is human" is more strikingly exemplified on the railroad than on any other mode of transit!   Why the fact of
defrauding a railway company should be so frequently regarded as pardonable, if not actually praiseworthy—as in the
cases of the non-payment of income tax—we cannot divine;
and yet there is almost as much swindling in the train as on
the race-course—and such mean swindling too !

There are many ways of endeavouring to evade payment,
such as collecting all the tickets in a compartment; when the
total is one short the culprit can very easily shift the blame to
another passenger, and the discovery of the deceiver is rendered
almost impossible.   People, thinking themselves overcharged,
take it out by "riding in a class of carriage superior to that for
which they have paid," and such revenge we presume must be
sweet or it would not be practised.   Until continuous communication through the train be adopted, as on the Swiss lines
for instance, no absolute check can be provided.   When this
is done, the vexatious delays for "tickets please," railway
frauds, and outrages of all kinds, will be stopped, and both
company and public will be gainers.

One very ingenious mode of being revenged upon a railway

company was practised by an Irishman. The story is not new, but it is worth re-telling.

Paddy had complained of having to pay more than a penny a mile on a certain line, and racked his brains as to how he could "pay out" the company. At last he hit upon an expedient.

"Now, boys, I have them!" he exclaimed triumphantly one morning as he entered the train. "The spalpeens! I've got them now!"

His aquaintances enquired by what means he had circumvented the officials.

"Bedad, I've done them," he replied in a confident whisper. "*I've taken a return ticket this mornin', and I don't mane to come back!*"

Distorted as was this ingenuity, another railway traveller succeeded in perfecting a swindle of such a plausible character that we may say that it almost deserved success. As related to the writer by a railway man, the tale is as follows,—

"After the passengers for Euston had entered the carriages of the last train at Holyhead, just as the train was starting, a very simple-looking traveller asked his opposite neighbour, a young woman, if all the tickets on the line were the same.

"'They tell me,' said the simple Irishman, 'that the tickets given to men an' women on this line do be different.'

"The young woman smilingly produced her ticket to Euston, to convince the silly fellow of his error. He inspected it, and handed it back with a shrug, and for the rest of the journey to Rugby kept silent. Here, our informant states, the tickets for Euston were then collected, and the simple one alighted. When he re-entered a compartment, not the same he had quitted, the watchful collector came, and demanded his ticket.

"'Sure you've got it, man!' was the reply.

"This the official denied, and as the Irish simpleton was positive, and the collector very decided, the dispute warmed. The would be traveller was detained; the train proceeded; the simple one stormed. The inspector came up; the

matter was explained. The Irishman described his ticket, and
mentioned the number of it. The tickets were examined, and
sure enough the pasteboard was found. The Irishman then
took the upper hand, declared that he would sue the company
for assault and battery, for detention, and expenses.

" The inspector was nonplussed. He could not deny the
evidence. The number and date of the ticket proved Paddy's
case. The traveller was interviewed, his demands were complied with. He remained at Rugby that night in comfort,
and was forwarded to Euston next day free of charge. He
preferred his claims for damages, and received a substantial
sum from the company in mitigation ; and all this because he
had noticed and remembered the number and date of the
ticket of his opposite neighbour in the boat train."

These are but specimens of the talent which is misused on
the railway. The issue of these pasteboard passes requires
quickness and good temper, with a knowledge of figures. The
price of the ticket is on the face of it generally, and on the case
from which it is drawn ; but sometimes several are required,
and addition must be rapid. The ticket next in succession is
pulled partly out by the dexterous middle finger of the clerk,
and is ready to hand in a second. The tickets are numbered
in succession, to 10,000 as a rule, and then recommenced ; the
dating is done when the ticket is issued.

Sometimes a partly faded and dusty first-class ticket comes to
hand. It has been exposed for some days, perhaps weeks,
since the last previous ticket of the class was issued. The
accounts are readily made up, because the intermediate numbers between the last noted and the ticket next to be issued
give the numbers of the tickets issued to passengers by any
train, or during any given time. The tickets, multiplied by
the price, give the sum for which the clerk is responsible.

The issue of tickets over other lines has given rise to the
establishment of the Railway Clearing House at which we must
glance for a few moments.

In the year 1842, when the capital invested in British Railways amounted to some fifty millions sterling, there was started, on the principal of the Bankers' Clearing House, a small establishment with half a dozen clerks, to clear railroad traffics, transits and tickets. The man who first conceived this idea was Mr. G. Carr Glynn, afterwards Lord Wolverton, who was at that time chairman of the London and Birmingham Railway.

Mr. Glynn took into his confidence a very able man, Mr. Kenneth Morrison, a retired Indian civil servant, who had a marvellous aptitude for figures. To him did Mr. Glynn turn, and not in vain, for Mr. Morrison soon got the idea into shape, and the Railway Clearing House was opened in Drummond Street, near Euston Square Station. Amongst the employés of the "clearing" have been Zachary Macaulay, Mr. Oliver, who has grown old in its service, and the late secretary, Mr. Philip W. Dawson.

To the late Sir James Allport has been attributed the first idea of this most useful establishment; but we believe that to Mr. Glynn the proposal is due, or if not the actual proposer— a position which some authorities have variously ascribed to Messrs. G. Stephenson and Morrison—Mr. Glynn was certainly the prime mover in the working of the Railway Clearing House.

From these small beginnings the present large, but by no means beautiful, establishment has arisen in Seymour Street. In 1892, this clearing-house had been in existence fifty years. Its business to a certain extent resembles that of the Bankers' Clearing House, which balances the amounts, the cheques, and decides what sums each bank is liable for; yet there is a very great difference in the mode of working. The Railway Clearing House constantly apportions the proportion of the cost of through tickets which each company, over whose line they were available, should receive; but it does much more than that.

The Bankers' Clearing House deals merely with cheques, and the balance either way is paid by a draft on the Bank of England, so no cash actually passes, but the Railroad institution, while dealing with the tickets, and the consequent proportion of fares, has also to deal with the immense number of transactions which arise from the continual interchange of rolling-stock: passenger coaches, trucks, waggons, etc., which are daily passing over lines other than those for which they were constructed.

Before the Great Western Railway altered its seven-foot gauge, passengers passing from its line had to change carriages; and such was the rivalry between companies that the public convenience and proper correspondence of trains was the very last thing to be considered. We have read how the South Eastern delighted to run a particularly slow, and continually stopping train immediately behind a Brighton express "Quite right," the ordinary observer would reply. But if the fast train happened to be a little late the slow train insisted on proceeding, and delayed the express all the way from Red Hill. On other occasions the South Eastern "express" would "dawdle" along in front of a fast South Coast train; and hundreds of travellers have had experience of this kind. But we believe a more enlightened policy now obtains.

The interchange of traffic from one line to another is looked after by the employés of the Clearing House. At junctions officers are in daily and nightly attendance to check the coaches, and note for the companies where and when they proceed. So with "goods" and coal-trucks and waggons. They are all noted; the numbers and the people to whom they belong are put down.

A certain time is allowed for return, according to the service in which the vehicle is engaged. Obviously, a passenger coach or even a train, will not be detained so long as a coal-truck, meat-van, or hay-waggon. The first-named is emptied and returned rapidly; the others must be shunted and unloaded; and after the time fixed has fully elapsed, a charge for detention, or demurrage, is made against the people or company who detain it.

These charges, at a certain rate, are debited and credited in the accounts in the ordinary way. This demurrage has been stated by a correspondent of the *Times* to amount to a period of detention equal to fifty thousand years, or just one thousand years per annum since the institution of the Railway Clearing House; which, multiplied by the number of days in the year, gives us a total of 365,000, or, in the fifty years, more than eighteen millions of days; and this again multiplied by the hours in the day, fixes the number of hours demurrage paid for since 1842 to 1892, to over four hundred and thirty-two millions! Those who like figures can easily ascertain the number of billions and millions of minutes which these represent.

When we consider the length of the mileage of the railways now open in England and Scotland, the extent of territory supervised by the agents of the Railway Clearing House will appear enormous. There are some twenty thousand miles of railway open. This comparatively small sum in no way represents the mileage run, for the London and North Western Railway alone represents a running mileage of 41,899,000 per annum. The total miles run by our railways, as near as one can calculate from available data a year old, is 313,470,000 which, if we call six days a week working days, gives us something over a million miles run in a day, to which England is by far the largest contributor, with about 843,000 miles; Scotland 120,000, and Ireland 42,000.

With so many hundreds of trains running so many thousands of miles, the number of passengers must, of course, be very large. We do not exaggerate when we state that a few thousands over eight hundred and fifty millions of people are annually carried on our railways! The odd thousands fluctuate a little perhaps, but the millions remain as a standing testimony to the British love of locomotion.

On the clerks of the Railway Clearing House, then, devolves the duty of looking after and checking the "through" traffics, and the tickets of this enormous number of travellers, carriages, vans, trucks and waggons. The receipts exceed twenty millions sterling per annum, from which the value of the

R

rolling-stock employed may be guessed at ! The Clearing House has to check the returns of merchandise as well as the tickets, and in the case of the former, there must, as in all store transactions, be a voucher of issue and a voucher of receipt. If these returns do not exactly tally, enquiry must be made, and the error or fraud discovered.

To conduct and carry out all these most responsible and arduous duties there are sixteen hundred clerks, under a managing committee, in Seymour Street, and about four hundred and fifty men employed in checking the traffic at the various junction stations. The clerks are, as in the Civil Service of the Crown, organised in divisions under superior clerks, heads of departments, assistant secretary, officers or principals, and the secretary. The managing committee, which consists of members of the different Railway Boards, meets quarterly, and again delegates seven gentlemen to superintend the actual working of the " House"; these are generally managers of the railroads or prominent directors, such as Lord Claud Hamilton and the late Sir James Allport of the Great Eastern and Midland respectively.

It would carry us out of our province were we to proceed to detail minutely the somewhat complicated business which is conducted in this remarkably well-ordered establishment. Enough has been said to interest the general reader in the efforts which are daily being made for his comfort and convenience by the Railway Clearing House.

COUPLED LOCOMOTIVE PASSENGER ENGINE, 1860.

CHAPTER VII.

RAILWAY RUNNING.—PUNCTUALITY AND THE PUBLIC.—
THE TIME-TABLES.

"PUNCTUALITY" we were informed at a very early age "is the soul of business," and if that be so we fear that Railway Corporations have sometimes very small business-souls. On all sides we hear complaints of unpunctuality, loss of other conveyance in consequence, loss of time, money, and temper. "For the life of me," exclaimed an elderly gentleman in our hearing, "I cannot conceive why these trains are almost invariably late! Come up when I may, my train is almost always behind time!"

This, we hope, was an exception; but there can be no doubt but that the Southern lines in our Great Britain are, as a rule, less punctual than the Northern lines. It is idle to make comparisons with the continental railways which go so slowly, and on which so few trains relatively run. It is useless for other reasons, one in particular being the indiscriminate issue of tickets up to the very last moment, and in many instances until after the last moment, advertised for the departure of the train from the terminus from which it starts. In this arrangement our continental neighbours may teach us a lesson, but whether we would profit by it is a question. On the Continent, booking offices are closed when the train appears, or even five minutes before its arrival! We have been refused baggage registration fully ten minutes before the advertised time of starting, and no tickets would have been issued to us—we had through tickets. On one occasion our baggage was thus registered at the Lyons Station in Paris, for Neuchâtel, by favour! But the platform inspector would not permit it to be placed in the van, though we were seated in the train; so we had to make up our mind to remain there, and proceed *sans bagage,* or turn out with our rugs and satchels. We chose the latter alternative, and had the pleasure of seeing the train start five minutes later. No doubt the pair of travellers in the compartment that we had vacated were pleased, though outwardly very sympathetic; but I can answer for it our little party were not so pleased at having to change the route, and proceed to Geneva an hour later.

Such an incident as this would be impossible in England. Tied as we are by red tape and County Councils, these bonds have not yet encircled railway travelling. But the fact is our English lines and station-accommodation are too crowded, and in some cases obviously insufficient for the traffic. The Victoria (Brighton) Station in London, and the South Western, extensive as they are, cannot be properly " exploited " in consequence of the narrow means of access. At present writing the delays in entering Waterloo are frequent. Perhaps, when the signals are settled, time will be kept; but so far, the chief effect apparent to a daily passenger is delay in starting and arrival, generally on the north side. There need be no want of punctuality on any line south of London if it be properly managed. When the Great Eastern manages to send crowded fast expresses from Yarmouth, Lowestoft, or Cromer into London, punctually to a moment, it is not creditable to the Southern railways to see the absurd South Eastern, Chatham, even Brighton and South Western trains, creeping in five, ten, and fifteen minutes late, often half an hour behind time.

Fewer trains and faster trains would suit. At certain hours it is well known some trains fill and some do not, and these "locals" do not much matter. But the long distance trains should be compelled to be punctual; the times should be fixed, with ample margins; and then the passengers, if not whirled along at 50 or 60 miles an hour, could, at least, count upon arriving at the time advertised, save in cases in which an accident had occurred.

But if we take the service time-tables and examine them, we shall not be so very much surprised at the non-punctuality of some lines, and we shall be pleased with the wondrous working of others—the Great Northern, for example. Such a number of trains are timed so closely behind each other, that unless every one concerned in its starting, running, and arrangements is exact and smart, the other trains will be set back. What does this involve, this punctuality?

Firstly, it means that the trains are marshalled, that is, made up, and backed into the various stations, wherever they may be on the line and its branches, in proper time. It means that the carriages shall be dusted, cleaned, oiled, coupled carefully, lamps ready, etc. That the engine is in trim, properly supplied with coal and water, and the engine-men steady and sober. To ensure the engine being in time the fire-lighter must have done his duty early, a sufficient quantity of water must have been in the boiler to make steam; the fireman must have been called, or at any rate have been present, early to see to the furnace; the driver must have examined his engine thoroughly before she quitted the "running shed," and he must quit it in good time because he has to run up, mayhap a mile or more, to the station, or across several lines, between trains and shunting waggons, and so pick his way to his train quite five minutes before starting time.

Then the men have to "couple up"—the fireman does this—and oil the engine, see to the fire and water, test the vacuum brakes, etc. The driver sees how many " coaches " he has on, so that he may estimate his load and run accordingly. The station-master must inspect the train; the porters must load up the luggage, passengers must be seated, doors shut, booking office closed, when the bell is rung. All these are actually necessary acts which must be performed ere the train can start; and then the signals must indicate the road clear !

This clear road supplied, the train starts; and supposing the "road clear" maintained, the driver and guard are primarily responsible for delays after. Of course the public will interfere at intermediate stations ; people will be halfway up a staircase as the train arrives, people will leap out of the carriages, and rush frantically, ere the train has pulled up, to the stairs, to save two minutes, as if they had not been home for years, and would not waste a moment when they reached it, or dawdle on the way to it. These delay intending travellers. Moral—provide separate exits and entrances. Many times has the writer seen persons lose a train in consequence of the block at the exit which is also the entrance gate to the platform. Here again the South Coast and South Western lines most particularly distinguish themselves—at Clapham Junction especially. When some girl or child is suffocated the Companies will wake up. The reason given to the writer for not making another exit is, " It would require another porter to take tickets ! "

Again, unpunctuality is caused by the engine-drivers reaching the several stations at the time stated for *departure* of the trains. On nearly every railway the time of departure is accepted as the time of arrival, and the public so accepts it. The South Western time-books say the time named is " that at which passengers may be sure of obtaining tickets " for the trains, and this may be true in one sense, but we venture to think that any one who trusts to that notice will be disappointed at Clapham Junction. The train is due away at 4.30 say; at 4.30 the booking office is besieged—in a tunnel. The staircase is blocked at the same time, for the train is just pulling into the station, and is already emptying! You have no chance! When you at length reach the platform the train has gone—time, 4.31½. You appeal to the ticket collector. 'Ah, you should 'a bin in time; 4.30's the time!" "I was here, but could not reach the train!" "Well, there's another at 5.42, and you'll have to change at Surbiton!" Consolation!

So much for trusting in simple faith to published promises that travellers will be certain to obtain their tickets and, presumably, places for the journeys at the advertised times at intermediate stations.

But it is easy to find fault; and with all the shortcomings of our railways it is sometimes marvellous what feats they do perform. The passage of a train demands the individual attention of more men than any other thing. Not only those actually in charge of it, but all the signal men, crossing men, and platelayers in its course must be on the alert. A wrong signal, an open gate, a loose rail, or chair, or sleeper, a stray animal, or wandering human being, a morbidly minded youth who wants to see an accident in the gloaming, and puts a sleeper across the rails : a dozen things may occur, any one of which, but for the incessant vigilance of the employés, would be fatal to the train. Still, while acknowledging all this, let the companies frame their time tables in view of the actual possibilities of traffic, and the public will soon fall into the change. They will, of course, patronise the fastest trains ; but when *all trains* beyond a certain radius are *equally fast*, they will not crowd one and delay it to the detriment of the others following. This equalization of pace of certain trains running to a distance, is the true secret of punctuality.

The number of journeys people perform now, when third-class carriages are attached to fast trains, is enormous. It has been calculated that every one in Great Britain takes a railway journey every fortnight, and these journeys account for six hundred millions of passengers annually, or a mileage run of about half that number of millions. The number of trains cannot accurately be determined; but they may be stated at an average of two trains per mile per hour!

To conduct these almost incessant trains, service tables are issued to the servants of the company by which the men concerned are guided. The engine-driver is informed not only at what moment he should stop at a certain station, or at all stations, but at what time he should pass signal boxes, junctions, level crossings, etc. Every train, no matter what it may be, is there put down, be it express, parliamentary, mail, goods, fast or slow, fish, mineral, or newspaper train! Besides these instructions special sheets are issued to warn all concerned as to alterations of line or signals, of which engine-drivers are more particularly warned. There are also "specials" to be provided for; and trains which run, or do not run "on Saturdays only," have to be reckoned with! Excursions come in order fairly well, but the Queen's train completely upsets all the traffic for hours.

Fogs, too, have to be reckoned with in the running of trains, and it is no unusual experience for a goods train to be detained for several hours while the passenger trains are creeping along. Under such circumstances the familiar landmarks are almost entirely obliterated; and the engineers must trust to their ears more than to their eyes on these occasions.

The time-tables are managed in a very simple but extremely ingenious way by diagrams on which each train is indicated by a thread of different hue, so that if the diagram be divided into hour spaces the positions and times of the trains will be seen with reference to the stations inserted beside the diagram. By doing this, and minutely subdividing the hour spaces, we shall see where one thread meets or crosses any other, and at what times the threads are opposite the stations, and when between the stations, at crossings, etc.

The printing of the tables is then proceeded with, and this is a very important operation. The least error would cause immense inconvenience to the public ; so the tables are most carefully read and revised by experienced men. Each company informs its rivals of the changes it proposes to make, and when the alterations are completed the sheets and books are printed. Advance proofs are of course forwarded to the publishers of time-tables as early as possible ; but sometimes they are not received sufficiently early to be published on the first day. "Bradshaw" is the principal purveyor of time tables, and we will look at this guide in our next chapter.

A LONDON AND NORTH WESTERN ENGINE, 1870.

ENTRANCE TO EUSTON STATION.

CHAPTER VIII.

"BRADSHAW" AND ITS HISTORY.—OLD TRAFFIC TABLES.—
REGULATIONS FOR TRAVELLERS.

IN the Romance of the Railroad this is a subject which cannot be passed over. "Bradshaw!" Look at it—a mass of figures which almost dazzles one! To some people this well-known volume is a problem which far surpasses the celebrated sixteen puzzle of a few years ago. But to any one interested in railways, Bradshaw is indeed a guide and a friend, if not a philosopher. Many pleasant journeys full of pleasant memories, or of equally pleasant anticipation, can be and have been performed by the inexpensive perusal of "Bradshaw." There is no difficulty in reading the figures, and to an ordinary intellect the book should be as easy as A B C.

Bradshaw, the initiator of the celebrated " Guide," was a Quaker, just as the man who invented the railway ticket was a Quaker, and the man who took up the first railway was also a Quaker. We owe much to our Friends. It seems that George Bradshaw was a publisher in a small way—of course he made plenty of money later—but in those days, say 1836, he was almost an anomaly—a publisher, and not wealthy! He first brought out a map of the roads, canals, and such means and lines of travel as then existed in England. He was assisted by the engineers—the great men then opening up the country; and the result was a connection with railway people when the new lines were introduced.

Mr. Bradshaw then conceived the idea of entering in a little pamphlet the times of the departure and arrival of trains. The pamphlet gradually developed into a volume, but it has been remarked as extraordinary, that the then existing railroad companies would not supply the information required by Bradshaw, who, not long after his development of the Guide, went into partnership with Mr. Adams, who at that time had already had considerable experience in the line of tourists' friend.

The price of the early numbers was sixpence, and the earliest extant are dated 1839. Mr. Maden, who investigated the subject, describes these volumes, which are in the Bodleian Library, as being bound in green and purple cloth, respectively; one for the Liverpool and Manchester, and the other for the London and Birmingham districts. The price of the latter was one shilling, a somewhat handsome volume gleaming in purple and gold like an Assyrian warrior.

Subsequently a " Companion " was published, also at a shilling, but ascending to the last obtainable evidence on the subject; the publication was intermittent, issued by fits and starts, in an uncertain and aggravating kind of way. To Mr. Adams is due the steady publication of the Guide, and it began its monthly career on December 1st, 1841. This is the date of the first copy in the British Museum, though, as we have seen, copies of other guides or companions were printed earlier.

"BRADSHAW'S RAILWAY COMPANION, containing the times of departure, fares, etc., of the Railways in England, and also hackney coach fares from the principal railway stations," with maps and plans, is before the writer. It is dated 1841, and priced at one shilling—a duodecimo of fifty pages. Compare it with the thick guide of the present date. There is a map of London in front, and a railway map at the end, containing the lines of railroad, all marked in red, which follow the course of the ancient roads. There are the London and Birmingham, and London and Southampton ; the Birmingham, Bristol and Thames Junction ; Birmingham and Gloucester ; Birmingham and Derby ; Manchester and Birmingham ; Leeds, Bolton, Bury ; South Eastern and Dover ; Taff Vale ; Great Western ; Great 'North of England ; Midland Counties ; Eastern Counties ; London and Brighton ; and many others now absorbed in the trunk lines, to the total number of 42. The Time-Tables were published in sheet at threepence.

The trains are marked as Mixed, Third-Class, Mail. The fast or first-class trains are timed to call at first-class stations only : such, on the London and Birmingham, as Tring, Wolverton, Weedon, Coventry. Then there is a " mixed " which calls at first-class places, which in its case are Tring, Wolverton, Blisworth, Weedon, Rugby. The Mail left at night, at 8.30 and 9.0 ; the former " mixed," reached Rugby at 11.58, the latter (Mail) at 12.15 stopping at the same stations. No smoking was permitted in the stations or carriages. Children under ten, half-price. The fares were variable, and were computed on the following basis :—

Four in carriage by day, or first-class six inside by night, to Watford, 5s. ; to Rugby, £1 4s. 6d.

First-class carriage, six inside by day, to Watford, 4s. 6d. to Rugby, £1 2s.

Second-class carriage, closed by night, to same stations, 4*s.* and 18*s.* 6*d.* respectively.

Second-class by day, 3*s.* and 15*s.*

The charges for a trip of 60 miles in these classes, respectively, were 17*s.* 6*d.*, 16*s.* ; 13*s.* 6*d.*, 10*s.* 6*d.* The third-class train took passengers' private carriages and horses to Birming-

BATH STATION IN 1845-1892.

Showing the old broad gauge carriages and locomotives.

ham for 14*s.*, £3, and £4 ; each with intermediate charges in proportion.

The Great Western appears in this pocket edition of Bradshaw as taking nearly three hours by the fast train, at 10.15 a.m., to reach Swindon ; the Cornish express occupies about 1 hour 27 minutes on that journey. To reach Bristol in those days entailed a journey of 4½ hours, which was not bad travelling, considering the eight halts, at Slough, Maidenhead, Reading, Steventon, Faringdon, Swindon, Chippenham, and Bath. The fares were clearly set forth, first, second, and third class ; to Bristol, 30*s.*, 21*s.*, and 12*s.* 6*d.* respectively ; to Reading, 8*s.*, 5*s.* 6*d.*, 3*s.* ; so the "parliamentary" fare had not been then ordained.

The South Western ran from Vauxhall, and its managers do not appear to have timed the trains for the public at way-stations. The trains were advertised to leave Vauxhall and Southampton for certain places on the way to the other termini respectively, but the intermediate stops are not clear. Thus the trains from Vauxhall are,—

To Southampton	. . .	7.0 a.m.
„ Weybridge.	. . .	8.15 „
„ Southampton	. . .	9.0 „
„ Woking Common	. .	10.15 „

and so on.

To Vauxhall,—

From Southampton .	. .	2.0 a.m.
„ Woking Common .	.	7.45 „
„ Weybridge	. .	10.15 „
„ Southampton, mixed	.	6.15 „

and so on.

The first-class trains took first-class passengers only, except a certain limited number of servants in livery, at 15*s.* The first-class fare to Southampton was 21*s.*

The London and Brighton trip was partly performed by coach. The trains left London, Tooley Street, at 9.30, 11.30, 1.30, and 5.0 ; they reached Hayward's Heath, 11.30, 1.30, 3.30, and 7.0, and thence by coach to Brighton in two hours more. Passengers were compelled to book to Brighton on the day preceding their journey, to ensure conveyance ; but they were not permitted to travel to Hayward's Heath by the coach from Brighton, unless proceeding thence by the trains. The fares to Brighton were 15s. and 11s. To Red Hill, from London, 5s. and 3s. 6d.

In like manner we have the North Midland, which, extending from Leeds to Derby, connected apparently with a line to Birmingham, and thence to London. In the London and Birmingham time-tables was a junction (9¾ miles from Birmingham) named Hampton, and trains are marked as in conjunction with the Birmingham and Derby Junction. So in those days the companies appear to have endeavoured to assist each other's passengers, instead of, as now, carefully timing their departures to *not* coincide with the arrivals of a rival company.

There was then also a Midland Counties line, which competed with the two last mentioned, and the rivalry at length culminated in such severe competition that passengers were carried through for a couple of shillings, about one third of the local distance fares on each line ; where rivals ceased from troubling, and directors were at rest. After some severe losses the three boards amalgamated, the line became one and indivisible, and from the three competitors emerged a " harmonious whole," which is now known as the " Midland."

The Great North of England appears in the early guide with a dozen trains, including both up and down traffic. The Mail, which left London at 9 p.m., reached York at 7.20 a.m., and Darlington at 9.20. Thence the journey was continued by coach to Newcastle. There were several other coaches in connection with the railroad which communicated with Northallerton, Ripon, Barnard Castle, Harrogate, etc.

There are numerous directions for passengers, scattered in notes in the margin of " Bradshaw," as to the care that should be exercised in noting the number of the carriage on which the luggage is placed, the notice that the " Manchester 7 a.m. train is the only one in which passengers can go to London in waggons "—a curious privilege. Passengers are also warned in this little but apparently complete guide, to be at the First-

Class stations five minutes, and at the Second Class stations ten minutes, before the time specified for departure. The passengers on the London and Birmingham Railway were requested to have their names and addresses *legibly written* ' on each part of their luggage," when it will be placed on the top of the coach in which they ride. The term "coach" is here preserved.

A regulation as to tickets provided that, as a rule, they must be used on the day of issue by any of the trains, at pleasure ; but under special circumstances a ticket could be exchanged for a new pass for the day required. This was a boon which the present-day director or superintendent would positively refuse under any circumstances, as he would decline to refund the money paid for an unused ticket, as experience can testify.

In those days, and indeed for many years afterwards, smoking was prohibited in the stations and carriages. Some railways now include the bye-law against smoking in the stations in their codes. The hackney cabs and Coaches in connection with the railroad made different charges. From Euston to the Charing Cross Station was two shillings in a coach, and one shilling and fourpence in a cab ; to Fitzroy Square, one shilling, and eightpence, were charged respectively ; while to Bedford Square, by Gower Street, was eighteenpence and a shilling ; the same journey through the Bedford Estate only cost one shilling in the coach, and eightpence in a cab.

The baby " Bradshaw," from which we have culled these details, is furnished with excellent maps of the several railway districts, with plans of towns, and of the lines. Cab-fares from all principal termini are given ; fares by the railway for all trains, and for people riding in their own carriages—these 2*d.* a mile—their servants riding outside, or in waggons ; and children, waggon-fare. There are many suggestive notes and directions in this little book, though for general information it cannot of course compete with our latter-day " companion," with which the public is so familiar, and of which the general reader knows so little.

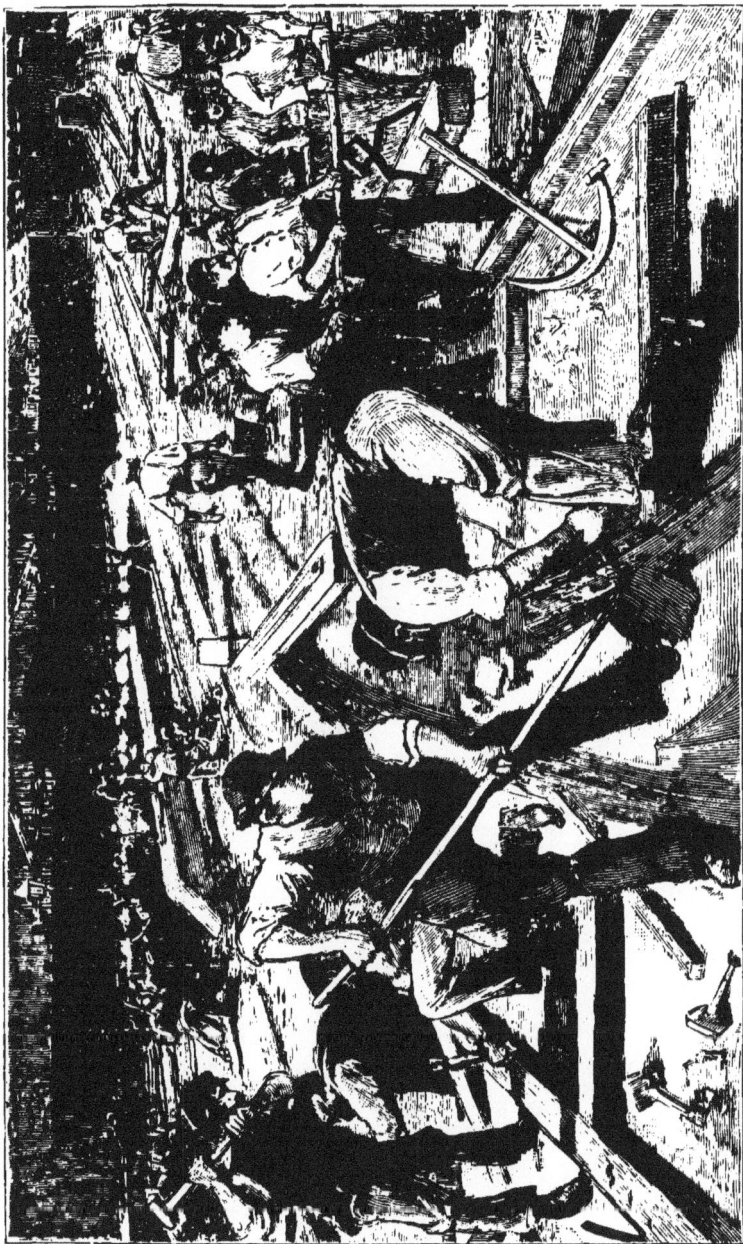

SCENE AT PLYMOUTH STATION DURING THE ALTERATION OF THE BROAD GAUGE LINE, 1892.

Why should not " Bradshaw " be taught, and the directions of the various lines of railway, their stations, the objects of interest in each, and their products, history, etc., be inculcated by means of maps? Many years ago the writer ventured to make this suggestion, but the gentleman to whom it was made only smiled. " It would only puzzle young people!" said he.

But the Civil Service Commissioners did not disdain to employ " Bradshaw " in their list of questions. One question which the writer had to reply to in competition was to the following effect. " Name the principal stations on the Great Northern Railway between London and York, their historical associations, etc., and the objects of interest to be seen in each town."

One competitor declared that he wrote against this question, " See ' Bradshaw.'" He did not pass! But the commissioners of that time recognised the usefulness of a knowledge of railway geography; and why should it not be taught in the High Schools and Board Schools of this era of over-education! Head masters please copy!

THE UP "DUTCHMAN" PASSING WORLE JUNCTION, SOMERSET,
AT 60 MILES AN HOUR.

Showing the complicated points of the formerly mixed gauges.
(From an instantaneous photograph by J. A. C. Branfill.)

CHAPTER IX.

THE BATTLE OF THE GAUGES.—WIDE *v.* NARROW.—THE CON-VERSION OF THE BROAD GAUGE.—CONCLUSION.

ANY history, however brief, of the English Railway, particularly of its more romantic aspects, would be incomplete without some notice of the famous "Battle of the Gauges," which raged in 1845 between Brunel and rival engineers. The debate in Parliament in June, 1846, was a memorable one.

Firstly, as to the gauge itself. Everyone knows that this term indicates the space between the rails of the system of tracks laid down on what is termed the "permanent" way, though it is certainly not permanent; it is continually being "picked" up and relaid.* However, the gauge is the space between the rails, which, in England, by the time this volume is printed, will be that generally adopted by all lines, although only lately by the Great Western Railway, viz. 4 feet, 8½ inches. In Ireland the gauge is wider, viz., 5 feet 3 inches. But there are in England, or rather in Wales, some narrow gauge lines, and the Continental and the American mountain railroads show "Tom Thumb" lines of very narrow gauge.

The English system of gauge, 4 ft. 8½ in., was taken from the cart track which, when measured for rails gave 5 feet, all told. But the breadth of the rails had to be included, and as each rail was 1¾ in. wide the inside measurement of the line was taken at 4 ft. 8½ in. Yet this did not find favour with every new railroad maker. George Stephenson's lines were all laid on such a base, but Mr. Brunel preferred 7 ft., the Eastern Counties, 5; the Caledonian, 5 ft. 6 in.; and, as already intimated, in Ireland a different gauge still. In fact, in Ireland half a dozen different widths obtained. The decision of Parliament in 1846 settled the gauge of Great Britain at 4 ft. 8½ in., and that for Ireland at 5 ft. 3 in. The Indian gauge is 5 ft. 6 in., which affords boiler-space in the engines; but the only reason we can find for making the Irish gauge 5 ft. 3 in. is the desire to do equal justice to every line in that distressful and peculiarly constituted country. All the gauges were added up, divided by the number of railways, and the authorised gauge is the result !

Stephenson, as we have said, adopted the "narrow gauge," and ridiculed the idea of the broad gauge, which Brunel, a man of the grandest conceptions and most magnificent ideas in engineering, initiated. Nothing less than 7 ft. would suit him on the railway. Nothing less than the *Great Western*, *Great Britain* and *Great Eastern* steamships would suit him

---

* It is called *permanent* to distinguish it from the narrow and ephemeral contractors' line, during construction.

on the ocean-way.   The Saltash bridge remains a witness
of his skill, and Maidenhead bridge of his daring brickwork.
The Clifton Suspension Bridge and the Thames Tunnel are also
associated with his name, as the atmospheric railway is with
his originality and decision of mind.

Yet, grand as were his ideas, he had to give way; and the
20th of May, 1892, saw the last broad-gauge train run upon
the Great Western Railway!   The *Flying Dutchman* is no
more; it has gone out of the system of the railway as completely as the "Phantom Ship" has passed from the surface of
the sea.

There is a feeling of sentiment and regret at this disappearance which we cannot help sharing.   We are sorry that the
"old Great Western" of our boyhood, its fine-looking locomotives, the *Balaklava*, *Swallow*, *Prometheus*, *Inkermann*,
Great Western, *Iron Duke*, and many others, on the 20th of
May disappeared as broad-gauge locomotives.   Our youthful
enjoyment and middle-age pleasures have been so often connected with the "seven-foot," that a regretful glance at the
deceased gauge is permissible.   Our very earliest romance was
bound up with the broad-gauge; our South Devon pic-nic
parties, our early inspection of engines, our first attack of the
"tender" passion, we may say, are all associated with G.W.R.,
and many a pleasant holiday in mature years has been begun
on the broad gauge!   R.I.P.

Brunel had his reasons for fixing on the broad-gauge.   He
considered that if his railway were of that width, and had
a sufficient number of branches, it would secure all the
traffic.   He thought, as other engineers also thought, that a
certain railway would occupy, feed, and be fed by, a certain
district.   At any rate Brunel's opinions prevailed.   His directors agreed with him; the rival lines, once intended to
meet at Willesden, could never meet again.   They had quar-
relled, and "strangers yet" they have decided to remain.
Brunel led his Great Western to Paddington; Robert Stephenson his London and Birmingham to Euston Square.   Such

lines could *never* meet and could never interfere ! So Brunel
fancied. . . .

It must be confessed that the first trials of the broad-gauge
were failures. The carriages were too short, the line laid on
piles, and baulks of timber with " bridge rails " jumped, and
shot the passengers from their seats. A complete change in
the rolling-stock and in the laying of the line on longitudinal
timbers was more successful. But these again had to be supported by stout planks between the baulks, and even then the
strain by the heavier trains was very great.

On June 4th, 1838, the line was opened from Paddington
to Maidenhead, and immediately considerable interest was
excited. Engineers were appointed as a commission to
examine and report upon the line, for which purpose facilities
were granted by the directors. Messrs. Wood and Hawkshaw
reported unfavourably, the directors replied, and the shareholders adopted their views by a large majority, but by no
means unanimously. The line prospered, and in June, 1841,
it was opened to Bristol.

The advantages claimed for the Great Western seemed to
have been secured. The speed was good, and " much exceeded
the rate of ordinary railway travelling " ; the " smoothness and
comfort " of the running are also noticed in the reports of the
directors ; and the public agreed with them. As regards the
speed, the Great Western trains, in 1845, travelled from London
to Exeter in 4½ hours—a distance of 194 miles ! It is not
much better in 1892.

There is no doubt that Brunel's pluck and determination
carried him through many difficulties. Nearly every engineer's
hand was against him, every newspaper argued the question,
and pronounced against him. His railway opponents increased.
Parliament decided the controversy, but though all other
lines took the narrow way as the safest, the Great Western
continued in the broad road, and began to perceive after the
lapse of years, as competitors multiplied, that it was leading
in the direction of destruction.

By the Gauge Act of 1846 the Great Western retained its 7 feet; but as traffic increased and passengers multiplied, as "through" arrangements became desirable, and the necessity for a "clearing house" for the interchanging traffic became evident, the narrow-gauge lines obtained the bulk of the business, they spread narrow-gauge arms into broad-gauge territory, and took the bread out of the mouth of the G.W.R.

Then the giant awoke and bestirred himself. He consented to have another rail placed on the track by which the narrow-gauge trains and waggons could run over his burly system. He mixed the gauge; and by degrees cast off the old rail which had so crippled him, and on his branches became a narrow-gauge. But the trunk remained firm. The storm of competition might break off a limb but could not affect the trunk; yet the decay proceeded. In South Wales the line was changed, until only about 400 out of a former total of 2,500 miles broad-gauge remained. The *Cornishman*, that fine 10.15 a.m. west country train, now runs narrow, and the original *Dutchman* and the *Zulu* will be no more seen.

The directors have long recognised the necessity for this reduction, and their policy is bearing fruit in the rise of the price of the stock. Less than a quarter of a century ago G.W.R. stock was as flat as "Eastern Counties'" was. Now it has overtaken and beaten the Great Eastern, and still keeps well ahead both in speed and price. But G.E.R. is a worthy rival, and will make its way ere long into a high position.

This, by the way, however. The Great Western directors had long been making their engines and coaches "convertible." When the clock struck the hour, and the workshop bell tolled the knell of the dead gauge, the transfer took place; in a few hours the carriages and engines were "converted" and turned into the narrow way with a speed which many people may envy. The engines of Sir D. Gooch have disappeared, and the smaller coupled wheeled "new express" narrow-gauge locomotives have taken their places.

The Queen's broad-gauge saloon has long been disused, and the narrow-gauge train now runs over the loop to Windsor.

As regards speed there can be at present little difference in the once rival system now dying out. The Great Western does not run so fast as the Great Northern ; and the " race to Edin burgh " proved that the North Western *could* do best of all ; but it does not ! In former days we remember the run from

LEAVING PLYMOUTH AFTER THE CONVERSION, MAY, 1892.

Didcot to London was considered remarkable in the time—53 miles in the hour ; yet it is an easy run, as any driver will tell you, being down hill when not level.

Now, even the South Western is treading upon the heels of its great rival, and the continuance of the narrow-gauge to Plymouth by this line from Exeter has been the last straw which has broken the gauge's broad back. It is gone—passed into the realms of the " has beens." Farewell, " broad-gauge " ;

thy place knows thee no more! Within 30 minutes the change of gauge was effected in each coach: the narrow-gauge bogie was substituted for the broad gauge by a most ingenious arrangement, worked by hydraulic power, by means of which the trucks, when released from the bodies of the coaches, were lowered to another level; the narrow-gauge trucks, being then raised up underneath the suspended carriage bodies, were fixed, and run out for use on the newly laid line! Thus the change was effected. Farewell, broad-gauge; thy life is passed; thy race is run! . . .

Born in 1838, the broad-gauge had grown up in 1845 to a lusty youth of 270 miles in length, and in 1867, to its prime of 1,450! Then a Royal Commission interfered, and denounced it—the Great Western had become almost isolated, and the existence of two gauges was firmly pronounced "a national evil." The remedy was drastic, but necessary. The gauge must be reduced by some means, and if the directors shrunk from killing it outright they had no objection to a "mixture."

So the "mixed line" was begun in 1854–6; the first narrow-gauge rails were laid in anticipation of the issue of the consultation. In 1862 the mixture had penetrated to the very head of the system; in 1875 it was carried on to the Bristol and Exeter section, on the line where the great. ten-foot tank-engines used to whirl the expresses on the road to Plymouth.

Still, in 1872, the change continued. South Wales was converted; and other extensions were, by degrees, treated in the same manner. In 1892, the main line remaining was made narrow, where narrow-gauge was not already laid "mixed." So from Exeter to Plymouth, Truro, Penzance, Torquay, and other western places, the broad-gauge has disappeared, and with it the big engines and the wide carriages. The broad gauge is a memory and nought beside!*

* The Conversion was completed on the 22nd May, 1892.

A DERBYSHIRE CAVERN.

THE ROMANCE OF THE SUBWAY.

CHAPTER I.

OLD PASSAGES AND TUNNELS.—MINING, AND MINING ADVENTURE.

SUBWAYS in Engineering include tunnels of various kinds: galleries, adits, and almost any underground passages constructed by human agency, whether for purposes of locomotion or for mining. There is, however, a difference between the tunnel and the covered way. The former is executed by driving a heading as in Railway Tunnels, and enlarging the excavation through the hill or bank. The latter is cut into from the surface, the way is opened beneath, and the earth is filled in again above the completed passage, as in the laying of pipes for gas, water, etc.

Tunnels are as old as the hills, and in some instances, doubtless, older. Nature with untiring hand drove many a heading into the cliffs by the sea; sent many a river under-

A RIVER IN A CAVE.

ground, and carved out many caves or subterranean passages long before Man, with his puny strength and imperfect stone implements, sought to dig his cave and tunnel his place of refuge. The derivation of the word "tunnel," or funnel, which has reference to it, is somewhat obscure; but *tonnelle*, an arbour, or tun (barrel), will supply a clue.

The underground passages formed by water and its wearing influences, are numerous; and, even if they are germane to our subject, are very well known. Visitors to America will seek the Mammoth Caves; visitors to Derbyshire will remember the Peak Caverns; those who have travelled in France may have seen that most curious and complete example of Nature's handiwork in the Dordogne; and other examples exist.

And living underground—under the snow—is still the habit of Laplanders. In the lamp-lighted dwelling, where the lights are always burning; where the fiddle or flute, the horn or drum, continue their crude music; where meals are movable feasts, and there is no count of time but as taste and fancy may dictate, the native manages to live happily. We associate " underground " with horror, a curious terror, perhaps, because it is with us inseparable from repellent darkness; but in the primitive subterranean house no such feelings exist. Darkness comes, like reading and writing, by nature; it is accepted calmly; and eating, drinking, dancing, chatting—and sleeping when tired of all these—follow each other, or exist simultaneously, without any fears, in Arctic dwellings.

This terror of darkness, which so many people not only suffer but entertain, is of course reasonable up to a certain point, for daylight is natural to us. But the objection to tunnels in the days of our early railways was great, though in a few cases tunnelling was strongly advocated. If any reader really wishes for the sensation of true darkness let him try the dark cell in one of H. M. Prisons. Let him be locked in alone by the gaoler, and left for five or ten minutes to cool. We will answer for it that he will not hanker for a repetition of the experience.

Of old, subterranean passages and hiding places found favour, and it could be shown that these excavations—secret underground passages, as well as quarries, mines, etc.—are of great antiquity. Regal tombs on this principle of tunnelling were not uncommon. We read of a king of Thebes who, as soon as he had ascended the throne, began to build his last resting-place, the entrance being by means of a long tunnel, which led to the death-chamber. The ancients were great engineers of tunnels and canals.

Long before the Christian era mining was carried on by the Phœnicians in Britain, or the "Tin Islands." Herodotus mentions the tunnel at Samos, which served as a aqueduct. A very interesting example of early tunnelling in the direction of canal or aqueduct, is the passage from Lake Fucinus to Lake Loris, constructed by Claudius Cæsar, and brought again to light in 1834–7, when the Neapolitans excavated the tunnel anew, and cleared it out. This subterranean water-way carried off the superabundant water of the upper lake.

Twenty thousand slaves were employed by Cæsar in cutting this aqueduct, and they worked for eleven years, shafts being sunk at intervals; and when the necessary level had been reached, the workmen cut sideways to meet their fellow-labourers. These shafts were sunk to a depth of four hundred feet: some perpendicular, others sloping down from afar. The gallery when excavated seemed to be some twenty feet in height, and an army of men were employed for several years in re-opening it. The surroundings and the tunnel were described by a writer in *Blackwood* in 1838.

It would appear that mining, and such underground labour, was considered very derogatory. Slaves, convicts, and the outside fringe of humanity were employed in this "infamous" occupation. In Siberia, to the present day, mining is considered the most degrading punishment, and the Goths were ridiculed in ancient times for mining [1]

Of course in olden days, when machinery and mining implements and assistants, such as gunpowder, were unknown, tunnelling was a very slow and lengthy operation. But civilization spread, free labour was employed, and science stepped into the tunnel; so now a passage can be cut in as many

[1] Tacitus.

months as had formerly occupied years. But in Tuscany, as in other spots in Europe, besides Paris and Rome, are miles of galleries communicating with vast chambers of the mines. Wonderful stalactites are discoverable, curious coins and implements have been brought to light. We may see the smoky niche in which the mediæval miner set his flaring candle or lamp, and guess at the primitive means employed to draw up the ore and soil by rope and bucket.

In other ancient mines the explorer will find the tools of bronze, even of wood and stone, and hart's horns, as well as the bony remains of man in his primitive form. In Sweden a body was thus discovered in 1719. It had been preserved by the deposits of sulphate of copper, and when found was quite recognisable. It was carried up to the surface, and made quite an exhibition of. People regarded it as a very ancient relic indeed, and might have continued so to regard it had not the well-preserved features been recognised by an old woman of great age.

Leaning over the corpse, she exclaimed, " It is he ! 'tis he ! " and fell senseless to the ground. When the spectators had recovered her, one of them asked her the reason for her alarm.

" Oh, 'tis he—Gustave, Gustave ! "

Now, no one knew Gustave, and further explanation became necessary. " Who is Gustave ? "

" He was my affianced," replied the old woman, weeping. I believed he had deserted me. Ah! I little thought that, while accusing him of faithlessness, he was lying down there, dead ! "

This is a very touching and pathetic incident. The young man, as M. Simonin proceeds to explain, had not been employed in the mine, but must have fallen into one of the many fissures in the ground. His tragic fate was unsuspected until many years after his disappearance. He remained apparently the same fair youth as then. She, a wrinkled, bent, old woman. He, just released—in death—from his grave ; she tottering to hers ; and here on the brink the lovers met again, after long years of silence, of doubt, and of suspicion! No doubt ere long they were for ever united in perfect happiness.

It appears that the general mining and tunnel-driving, in the "owld ancient times," were for metals and for aqueducts. The use of coal, with which most of our present-day mining is associated, did not begin till a comparatively late period of history. The reader will easily arrive at the period at which coal began to be a household adjunct if he search the records of the thirteenth century, when the mineral is first mentioned.

The "curfew" did not apparently apply to coal in the form in which we have it. The peat (or as it is called in Ireland, "turf") or wood was burned in the wide hearths and beneath the spacious chimneys. The "hearth" on which King Alfred burned his cakes and his fingers was most probably nothing more than a hollow in the floor in which wood was burned, and covered up when it became necessary to "put it out." The various illustrations depicting Alfred at a chimney, and baking cakes on the hearth-stone, must we think owe something to the imagination of the artist, and the license of the rhymster even of extravaganza is not quite admissible.[1]

But this is not history, nor tunnelling. The Irish cabin still reveals the primitive hearth; the curling turf smoke, which makes the uninitiated cough and feel ill, but which cures the accustomed bacon, is not altogether unpleasant; but the sulphurous fumes of coal—cheap coal too—would be most irritating and quite out of the question in olden time.

In Mr. Galloway's "History of Coal Mining" we come across the date on which coal seems to have been first used, He says:—

"As early as the year 1228, a lane in a suburb of the metropolis is mentioned under the name of 'Sacoles Lane,'"—*i.e.*, Sea Coals Lane—clearly showing that some trade in coal was going on there. "This lane," he adds, "was also called 'Lime-burners' Lane'; and it is well known that the burning of

---

[1] "Half *black'd* the *stove* !
And I'm *'Alf-red* the *great* !"

lime was one of the earliest uses to which mineral coal was applied."

Here then we have the date of the introduction of coal as 1228, or thereabouts ; and up to the commencement of the seventeenth century coal was not obtained from deep workings. Moreover, it was disliked as fuel, although gaining ground in public favour, and most profitable to the purveyors, as it is to this day. Henry III., in 1259, granted a charter "to dig for cole."

It would be very interesting to follow the course of the advance of coal in public estimation, but we cannot do so here. The history of the production of coal is a very sensational one, and life underground is one not to be envied. The carrying of the sea-coal became an important industry. Thousands of ships were employed in it. Coal was applied to smelting iron ; and the manner of the discovery of coal in Belgium in the twelfth century is legendary.

Coal in the seventeenth century was merely tolerated ; it was used under protest, and with extreme reluctance at times. In the year 1773, however, there were thirteen colleries on the Tyne ; and the positions of the coal-fields near the three rivers —Wear, Tees and Tyne—gave them, as it gives them still, great facilities of carriage for "sea-borne" coal. From the thirty-four colleries or so at work at the commencement of this century, there have sprung into existence nearly three hundred ! In Northumberland the railroad was first developed, and the locomotive drew the wagons in Cornwall or on the Tyne-bank. Trevithick and William Chapman found their opportunities, but lost them ; yet Chapman's locomotive, with its indented "scored" wheels, lay neglected in the rope-walk he possessed, long ere the *Rocket* aroused the admiration of the spectators at Rainhill.

Some of these colleries are now famous, and "familiar in our mouths as household words." What is more in common use than "Wallsend" Coal? And yet there is no such thing, strictly speaking. That celebrated mine is drowned, and remains a memory only. Perhaps some readers may not know the reason for the name of this celebrated colliery. It was so called from the Roman wall at the end of which the mine was opened in the eighteenth century, but it was drowned at last. Necessity then became the mother of the invention of the engineer ; and the working mechanic, the practical man, as we have seen, came to the rescue. The steam-engine was erected ; the water was pumped out of mines ; coal-trucks had to run as lightly and easily as possible —we know the rest. The fathers of the steam-engine, the railway, and the locomotive came forward, in the persons of Savery, Newcomen, Watt, Trevithick, Hedley, Stephenson and others, who have been the benefactors of mankind.

The pumping engine, which James Watt introduced, was, and practically the same machine is, a great boon to the miner ; the direct-acting pump being an improvement on the original machine. Notwithstanding all the care exercised, water sometimes gets the upper hand, and the levels, galleries, and drifts below are flooded, and ofttimes lives are lost.

There is one very serious and dangerous feature in mining operations, which is the upheaval of the flooring of the drift or level in consequence of the pressure of the roof! This may at first sight appear impossible, for one will naturally inquire how the *roof* can affect the *floor*. In every coal-mine there are pillars of coal left standing when the surrounding mineral is cut away. These pillars support the roof, but the tremendous weight of the upper strata presses or crushes the coal-pillars into the soft flooring, and by degrees the roof there settles down until the resistance is sufficient to support the pressure. This crushing together proceeds slowly, and is called "The Creeps."

SETTING A TREE IN A MINE.

Some of the levels or passages in the mines are shored up, or timbered; and along these the trollies run, sometimes drawn by ponies, or sometimes travelling by their own momentum. In former times young lads used to be employed to drag these coal "skeps," and terrible experiences they had.

The interior of a mine has a fearful fascination for some people, but we are unable to share the wishes of those who want to "go down again—as often as possible."

The descent into the underground passages is, of course, made by the pit-shaft, and means of descent vary. Let us go down.

T

SURVEYING A MINE UNDER DIFFICULTIES.

CHAPTER II.

ACCIDENTS IN MINING TUNNELS.—COAL AND ITS COST.—THE WORKING OF THE MINE.

THE method generally employed to descend into a mine is the cage, which runs in guides. There are two cages—one going up full, the other coming down empty—in the separated shafts, which are funnels or perpendicular tunnels leading to the underground workings.

There were, and in places there still are, other means of ascending and descending the shafts—buckets, and even a chain with cross-pieces, or a loop in the rope in which the miners place their feet, hanging on by their hands, lamps de.

274

pending from their waists, giving them the appearance of an animated chandelier. Many anecdotes have been related of the dangers incurred in days past by collisions and other accidents which occurred in the shafts. If the miner is reckless, he is also brave, and has often preserved his life by readiness and presence of mind.

On one occasion the "corf," or basket, in its descent came in contact with some tubs ascending. The chain of the basket snapped ; the men had only just time to catch at the ascending tubs and hang on while they were slowly wound up to earth again, in a most perilous position, pendent in the shaft.

On another occasion an overseer was nearly jerked out of the tub. He bent his leg as he fell, and hooked himself to the tub, his head and body hanging over the chasm. In this way he was carried up ; but rescued on the way, before he had reached the top.

But now-a-days there is not this danger, and the visitor, enveloped in proper dress, a lamp in his hand—mayhap a candle in his hat—may descend in safety. The descent is quickly accomplished, and the wonders of these tunnels and underground passages—the work of the mining engineer and the miner—will surprise the stranger.

The subway here reaches its greatest development. Arrived at the bottom of the shaft the numerous ways will be seen branching off, the galleries extending many miles under the surface of the earth, and in many cases beneath the sea : an immense town with migratory inhabitants who live and work down here several hours a day, taking turns, or "shifts"—men and boys of all ages, and in some mines (on the Continent) women and girls.

Standing at the bottom of the shaft, the roads strike out far into the seams. They are mapped out (surveyed), named, and as well known to the pitman as London was to Samuel Weller. Cross-roads, squares, and alleys are here, up along which the coal is pushed in the direction of the shaft. All through the twenty-four hours this underground city is awake, and lighted with lamps and candles; it has its tramroads and its horses, even its locomotives, but of the display of up-ground towns is there none—all is black.

Through these many miles of subway you may wander with a guide. You will reach a door which is shut; a lad will open it when the train of "rolleys" comes along, and close it when the train has passed. These lads are called "trappers"; they have to sit still, often in darkness, and open and shut the doors for the passing wagons. The doors are used to preserve the ventilation, and keep the air circulating in its proper path and direction.

It is in the matter of survey and planning of the various galleries that the mining-engineer can display his skill. He must make his surveys by lamp-light, and yet get his levels accurately, for it is essential that the tunnellings must unite with the other planned ways. The coal is picked out by various methods; sometimes the thick pillars remain standing, and the whole mine is honeycombed.

In some of the principal thorough-fares men can walk erect, and as these are fairly well-lighted the danger is reduced to a minimum;

SEARCHING FOR FIRE-DAMP.

but in some of the smaller tunnels the heat is great, and progression not so easy, nor is working very convenient. Lying on his back or on his side, the sturdy miner, his dim lamp beside him, hews and picks out the black diamonds for our use. In some mines machinery is employed for digging, on the principle on which the Alpine tunnels were excavated, by means of electric force or by compressed air.

But as a rule the diggers are men who load the "rolleys" which are driven along the rails to the bottom of the shaft. The quantity raised by the powerful machinery now in use may, and frequently does, amount to a thousand tons a day, when the men are not "at play."

The miner has many enemies in his calling. Of these fire and gas, choke-damp and fire-damp, are the worst, though water is also a serious foe. Water rapidly accumulates within the workings of the mine, and may remain behind its rocky barrier until some unlucky tap sets it free. The sea has actually been tapped in this fashion, and has percolated through the tiny aperture in the rock until a plug has kept it at bay.

Down, down beneath the deep these long workings extend many hundreds of feet under the ocean, whether you measure the distance perpendicularly or horizontally. Overhead the sea swells and tosses its waves; the dull roar may be heard at times in the mine. Even the rattling of the pebbles has been heard in the underground passage, so dim, so dismal, and so dangerous. There would be no escape if the water once made its way through the few feet still remaining between the summit of the subway and the bottom of the sea.

Men working in these submarine passages have been terrified by the roaring of the sea, and have decamped in fear. And indeed it is a fearsome thing to stand beneath the sea in its rage as it dashes the pebbles angrily upon the beach over your head, only a few feet of rocky bed existing between you and death!

Yet the miner works there; and in one celebrated Cornish mine was a blind man, who knew his way so thoroughly about the many passages and galleries that he became a guide to the workmen, thus rivalling Jack of Knaresborough, of whom we have already written. When the guttering candle had gone out, the blind man would find his way through the tortuous subways, across the chasms bridged by shaking planks, through mire and puddles, to the shaft and the blessed daylight which he could never see again.

Fire is the foe which the miner has to guard against. It may be produced in the course of blasting operations or by spontaneous combustion. At times a mine is full of fire, and is "hermetically

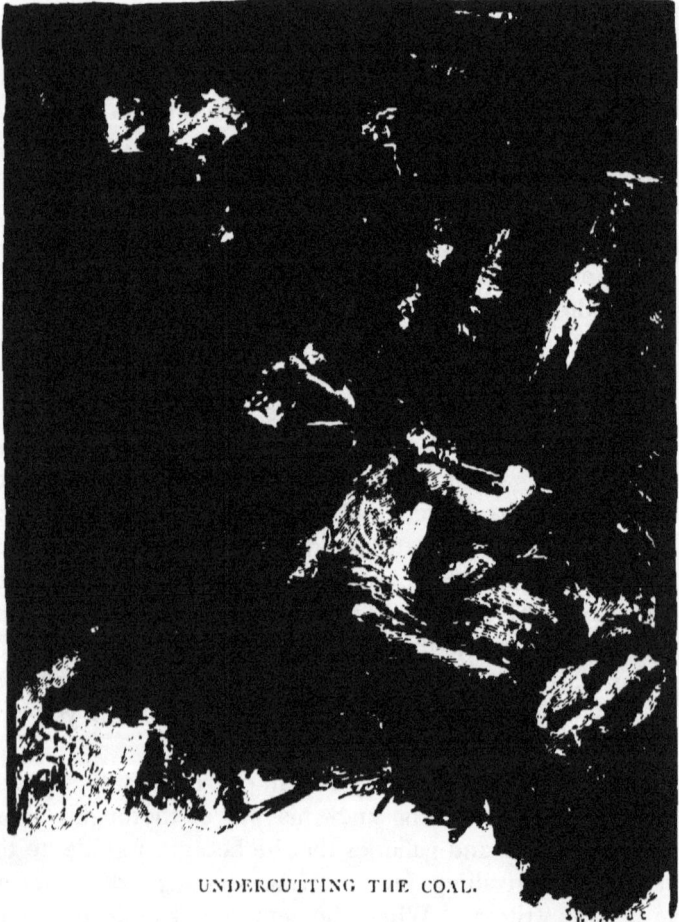

UNDERCUTTING THE COAL.

sealed," until no air can possibly penetrate into the workings to feed the flames. This successfully accomplished, the mine may be reopened after a while; but cases have occurred in which by some undiscovered aperture air has entered, and then the coal continues to burn until "drowned out." It may happen, as has happened in St. Etienne, that the coal has been burning for a period beyond the memory of the oldest inhabitant ; and the surface is baked, while the vapours given out remind the spectator of a volcano.

Similar instances have occurred in England. Staffordshire could furnish many examples; and while the subways were on fire the earth warmed and kindly supported an almost tropical garden on which no snow ever lay. These fires are checked by walls and partitions, and are not regarded as deadly accidents ; it is the explosive gas, when it meets with the atmospheric air, which is the cause of so many deaths—fire-damp !

There is no warning here. A sudden lightning flash, a roar of thunder along the galleries, strikes the miners down, and announces their death almost simultaneously.

There is hardly time to escape. The pursuing flame may overtake the fugitives,—"a roaring whirlwind of flaming air,"—tears away every obstacle in its path ; and the miner is thrown aside like a puny toy, burned to death, or mayhap only scorched, bruised, battered, or buried alive until timely rescue comes.

Death generally stalks unchecked through the mine on such terrible occasions. Choke-damp infests the workings—that terrible "after-damp" which so surely suffocates the half-insensible hewer, the little "trapper," or the "putter" in the gallery, and seizes its victims cruelly, relentlessly.

Such scenes are not uncommon, unfortunately, even now ; but before George Stephenson and Sir Humphrey Davy gave the mining world their "Geordie" and their Davy lamps, explosions were very frequent. Heartrending scenes were witnessed : the wrecked props and brattices ; the fallen blocks of mineral ; the dead men and horses ; the boys and ponies lying in all sorts of attitudes "asleep"; a few maimed and bruised miners groaning in agony, unable to move. Many widows, and orphans, and girls, insane from grief, remained above to bear sad witness to the catastrophe.

Before the time of George Stephenson mines were lighted by wheels of steel, which revolving rapidly threw out sparks from the flint they struck in their revolutions.   This crude and

WAITING FOR THE BLAST.

uncertain method was the invention of one Spedding, of White-haven, who carried the inflammable gas to the surface in pipes, as we now conduct sewer gas over our roofs.   He even proposed to utilise it for lighting purposes in the streets, and in general mining operations proved himself a skilled engineer.

Of course the "steel-mill" was only used in places where naked lights were infallibly unsafe. But even the mill sometimes caused an explosion. Thoughtful men endeavoured to find some substitute for the dangerous candle, and the insufficient steel-mill, whose sparks gave such "glimmering light." To a Doctor Clanny the first miner's lamp is properly attributable ; and his "Steady Light in Coal-Mines" attracted the attention of the Royal Society.

But this lamp did not prove a success. The promoters of the desired improvement then applied to Sir Humphrey Davy, who made numerous experiments on his return from the Continent. He saw the miners, and called upon Dr. Clanny, who showed him the lamp he had invented. When Sir Humphry Davy returned to London, he discovered "that explosive mixtures of mine damp" would not pass through small apertures or tubes ; and that "if a lamp be made air-tight at the sides, and furnished with apertures to admit the air," it would not "communicate flame to the outward atmosphere." This was in 1815, and on the 9th November of that year Sir H. Davy read a paper before the DAVY'S SAFETY LAMP. Royal Society on the subject.

Meantime George Stephenson had been making certain experiments at Killingworth in the same direction, and had also produced a lamp which gave light but did not explode the gas. However, although the priority of invention has been claimed for Stephenson, the safety lamp is no doubt to be attributed to Sir H. Davy, who at the end of 1815

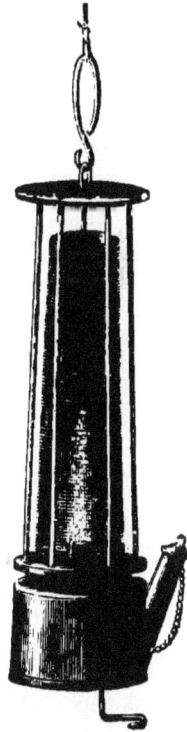

produced the lamp of gauze-wire so fine that no flame could penetrate it.

This most valuable invention put an end to the crude and dangerous means till then employed to get rid of the gas by firing it. This was done to free the workings, and the man who undertook the dangerous duty was called the "fireman" in England. In France he was known as the "penitent."

Attired like a monk, masked, and protected as far as possible from fire, the devotee proceeded to the place where the gas was "blowing." By means of a candle fixed upon an elongated stick he managed to fire the noxious mixture, and by crawling along the ground he escaped the fumes that rose from the explosion.

Electric lamps are now employed in some mines. Many other varieties of the safety lamp have been used in "fiery" mines, where explosions still take place, owing to the curiosity or recklessness of the miners, who want to see how the lamp acts, or who, in defiance of regulations, *will* light a pipe and blow themselves, their fellow-workmen, lamps, pipe, and all, in fragments, to eternity.

DAVY LAMP (SECTION).

Underground life, in all circumstances, is attended by many other dangers, such as the falling in of the roof or the irruption of water, which sometimes breaks through the workings.

Instances have occurred in which rivers have broken into the mines and drowned the men. The Tamar behaved thus. Resenting the disturbance of her bed, Tamar, the giant's

daughter of the Moor, had her revenge and flooded a mine which had been opened beneath it.

Mr. Dunn relates the terrible flooding of the Heaton colliery in 1815, when the imprisoned waters were accidentally released from the old pits, and rushed with terrible impetuosity through the newer excavations. No less than ninety human beings were imprisoned and drowned, the bodies being, when eventually recovered, soft and pliable as clay. The sensation: of the poor fellows may be imagined. Cut off from all hope of rescue, they must have died a thousands deaths in dying.

Nor are these terrible inundations at all uncommon. The underground pioneer is at all times liable to death by drowning. The dangers of the pits are well exemplified by the accident in the collieries of the Loire, where the men suddenly tapped the hidden reservoir. A narrow wall of rock only had divided them from the stored up waters. They poured through and descended in a resistless flood upon the miners. The men rapidly re-treated into another of the galleries, but unfortunately it had no outlet—it was a *cul de sac* ! They crowded up at the farther end, contemplating the rising waters with terror. No help arrived, though in such cases we know no efforts are spared to reach the imprisoned, the dying, or the dead.

A consultation was held on the bank, the well-drawn plans were examined, and those above ground very quickly became aware of the position of their unfortunate comrades. Not only this, but the depth beneath the surface was ascertained, and the direction in which the workers had sought safety was known. But how could even this knowledge avail when the mine was flooded ? To enter the mine by the ordinary passages was impossible. A new tunnel must be made. This was the only chance of saving the lives of the men standing on the brink of the flood-water, waiting in darkness and silence, in hunger and weariness, for rescue.

Promptly it was decided to bore and cut down in the direction of the upper end of the sloping gallery in which the miners were. The rescuers worked incessantly ; days passed, but no

response came to the ringing blows of pick and spade which
the earth must have communicated to the prisoners. Were
they *all* dead?

At length a joyful sound was heard. Men's voices came up
in muffled accents through the willing earth, and the labour of
the rescuers was doubled. A boring rod was hammered in as
far as possible to reach the gallery, and to admit air and light
when it had been withdrawn. The borer penetrated the last

AN EXPLOSION AT A COLLIERY.

stratum; it was pulled out, and, joyful to relate, the cry came
up, "Give us light first—light before food!"

The iron was withdrawn, the hole was enlarged; a ray of
heaven's beams was sent into the black hole in which the men
had lived, literally "in darkness and in the shadow of death,"
for six days, with no sustenance save their candles and their
leathern straps, which they had greedily devoured. The condition of the unfortunate men in the drowned mine may be
imagined.

This is a sufficiently miserable tale, perhaps, but the annals of Mining and the true records of the "winning" of coal supply us with many more startling adventures. Sometimes the inconsiderate act of a miner will plunge the entire mine under water, and imprison the men for days, even if they are left alive. In other places' Nature will suddenly cause a terrible catastrophe such as happened in 1862 on the continent of Europe. This was it :—

It seems that a sudden and violent storm burst over the district in which the mine was situated, and the neighbouring streams, which, save in winter, scarce rose to the level of a brook, were rapidly swollen by what some residents termed a water-spout. This cloud-burst filled the beds of the streams, they overflowed, and the water, deflected by the giving way of the bank, rushed into a fissure in the ground which opened to the roof of the mine.

Tons of muddy water rushed down this gap ; the roaring within the mine alarmed the occupants, who could not account for it. They very soon realized the situation, however, and sought means of escape. A few did succeed in escaping, assisted by the self-devotion of one of their number, who again and again descended the shaft in a tub, and explored the workings as far as possible, dragging half lifeless forms to the air and light, and sending them up in his bucket. He continued his work until the waters swept all before them, and the workings became filled ; the depths of the mine represented a lake, while many explosions testified to the terrific upward pressure of the air.

Assistance was speedily at hand, but four-and-twenty hours elapsed before any one could reach any spot where, judging by the carefully made surveys and plans of the interior, the men were supposed to be, if alive. While signalling by knocks the rescuers were delighted to hear distinct responses of the same character, and arrangements were immediately made to reach the men. But some twenty feet of coal wall intervened. How was it possible to pick out a passage in this barrier in the available time? Almost impossible if the days were to be counted by weeks; the cut might be accomplished within a month, but when hours were precious the labour to be done seemed useless to contemplate. However, the miners did not rest to think,—they acted. . . .

At five different places drifts were begun. Men worked like heroes, like Titans, one at a time in the close dark hole, in heat almost unendurable, picking as hard as arms, and hands, and pick could move and remove the hard coal, which was rapidly carried off. No cessation of this high pressure took place day or night, and at length a sound was heard. Voices penetrated the barrier, which seemed to be harder than ever.

At midnight, dark midnight outside, and still darker night within, with hardly any air, and dying lamps, access was gained to the small hole in which three miserable miners had been expecting death. They had avoided the flood, but the partitions had given way; a wall of solid coal had fallen down and imprisoned them—buried them alive. Two were found living still, but the oldest of the party had succumbed.

Other men and a lad were after many days rescued. The sufferings they had all endured were fearful and heartrending. They had all rushed away in advance of the flood, and had been driven higher and farther up the galleries, the water pursuing them with deadly persistency. Their lamps had been extinguished, the pressure of the air was very great, causing a singing or buzzing and a tightness in the ears, which was both disagreeable and painful.

The water, still very deep, imprisoned them; and they feared to sleep, lest they should roll into the lake below. They scarcely dared to drink at times; one man braved the chances and drank, but his parched lips came in contact with a corpse, and the horror of the situation drove him mad for a while. Others were rescued in a pitiable condition, but out of one hundred and ten pitmen only five absolutely recovered.

But even this very small result could not have been arrived at if the mining engineer and surveyor did not make plans so correctly. By their skill the whereabouts of the pitmen or the miners can be ascertained, and the diggers may be rescued.

The incidents above related are some of those to which the worker in the mining subways is liable. He " wins " the coal, after a hard struggle sometimes. It is on record that boring was continued for four years at Gosforth Colliery, when the coal was won at a depth of eleven hundred feet. To celebrate this triumph, a grand underground ball was given in the pit, in an excavated space flagged and arranged for the event, and brilliantly illuminated.

This dance took place during the morning and afternoon. The guests began to arrive as early as ten o'clock, and continued to arrive for three hours. Workmen, pitmen, their wives and daughters, gentlefolk, ladies and local magistrates united in this new ball-room, wherein the dance lasted until 3 p.m., thus reversing the usual surface hours for such entertainments. The company returned in baskets, new and carefully lined, to the upper regions.

The pit then ready was worked in the manner already mentioned, pillars and chambers being left excavated in the coal. This mode is termed " post and stall." Long-wall working is hewing progressively at the seam, leaving no pillars, but substituting props and struts of timber. Of course, the latter method is in deep and extensive collieries somewhat unsafe, for the roads must be kept clear, and natural support is the best. Some pits extend their galleries over hundreds of acres beneath the surface, and more than a hundred miles of cuttings or subways are not uncommon in mines. The pillars are some thirty-six feet by seventy-two ; and as little as one-sixth of a mine has been worked, the remaining five-sixths of available material being left to support the roof.

Even with this immense proportion the safety of the pit is not secured. The "creep" may set in, as already stated, and this movement can only be checked by erecting barriers. The coal remaining is won by digging into the soil or rubbish propping the roof, and when the mineral has been extracted in dangerous positions, the props are taken away—a very risky business.

We will conclude our glimpse at the mining subways with a few words respecting the winning and winnowing of the coal which is brought to bank.   The black diamonds in their setting of basket or " corves " are screened through immense sieves.   This business is known as " teeming " in some places, and "teeming rain," or rapidly dropping rain, is a term we have often heard.   To " set the teems on fire " is to be energetic in screening, and by causing friction by energy burn the sieve in your hands.   But coal " teeming " is done by tons through traps perhaps.   The largest coal, screened through a ⅜ seive, is called " wallsend," " seconds " and " nuts " come next in order.   So, as a writer has pertinently remarked, the term " wallsend " is now applied to the *size* of coal, instead of to the *place* it comes from."

There are many little secrets connected with the coal trade into which it is not our business to pry.   But one puzzle is—how is it that the coal-mines are always " working at a loss " and yet thriving ?   " We live on losses," said a partner to the writer one day ; and so we must suppose they have found the secret of making money out of nothing, and have discovered the philosopher's stone—COAL.

MOUTH OF MONT CENIS TUNNEL, AT BARDONNÊCHE.

CHAPTER III.

THE ALPS.—THE MONT CENIS, ITS PASS AND TUNNEL.—THE
STORY OF SÒMMELIER AND HIS MATES.

NO ground is more traversed in the present day than the Alpine regions of Savoy, Switzerland and Italy; and yet how many of those who travel in those countries can give a clear and intelligible account of the scenery through which they are too often hurried? Now-a-days, when the train rushes with us through the Alpine passes, or valleys, we frequently see the traveller, Tauchnitz volumes in hand, peacefully perusing the latest work of his or her favourite author, or calmly sleeping the pleasant hours away.

But even if we are more fortunate in our companions they have scarcely time to investigate the incidents connected with the wonderful and romantic engineering works in the Alps. There they are—that is sufficient! The mountain railroad, the steamboats and tunnels are there : and are merely railroads, boats, and holes in the ground—but no more. Of their surroundings, history, and the means by which, the men by whom, they were constructed, the toil, danger, and romantic details connected with these stupendous works, the ordinary tourist cares little. But the information is by no means "dry." The narratives have been related by engineers twenty years ago ; but who now reads descriptions twenty years old?

Nevertheless, they are worth looking at; and these marvellous engineering feats are always worth recording, particularly when one has watched the annual development of the various projects—has visited the sites and seen their ultimate success, during a space of nearly thirty years, before the great tunnels were made, in the days of the *diligence* and the sleigh.

The Alps, as a chain of mountains, are as familiar to us as— even more familiar than—the " Hog's Back." At school we made the acquaintance of Hannibal, who crossed them, and of Napoleon, who constructed three out of the five most important roads which traverse the Alps—viz. the Cenis, the Simplon, and the St. Bernard routes.

Not far from the centre of the north-western curve of the chain of the Alps rises the Mont Cenis, from which the first great Alpine tunnel takes its name. Why it is so called we are not able to state. The tunnel does not penetrate the Mont Cenis at all, and, as a matter of fact, that mountain is not within sixteen miles of either entrance of the tunnel, which pierces Mont Fréjus, the Grand Vallon, and a height on the Italian side known as the "Col de la Roue."

Having thus cleared the ground, we may look around us and see whereabouts we are. The traveller to Turin will note as he goes, in the daylight, from Aix-les-Bains and the Lake of Bourget to St. Michel, the river Arc, which runs into the Isère, and with it swells the Rhône. The Arc rises near Mont Cenis, the principal mountain in the district, and the ·valley watered by this stream is paralleled by that traversed by the Dora, the Alps rising in all their majesty between them, and effectually separating them.

Well, what of it? readers may ask. Thus much : The rivers approach each other, and in their progress have in the course of centuries come as near as possible. So the valleys of these streams were selected as the best sites for the two entrances to the proposed tunnel, when the question of uniting Italy and France by a subterranean bond was seriously discussed.

The Mont Cenis Pass is an ancient one, for King Pepin crossed it with his troops in 755 A.D., and Charlemagne followed his example, as a good father should be followed ; and the hospice—if it be still standing—owes its existence to the Great Charles. Still, although the pass existed so many hundred years ago, it was by no means a road. Even Hanni-bal seems to have rejected it (at any rate, he passed it by as unsafe, and crossed by the Little Bernard route) and it was reserved for Napoleon the First to cut the carriage road, which runs a distance of fifty miles—a splendid feat of engineering—between St. Michel and Susa.

This was sufficient for a while ; but some thirty years later Signor Medail, of Bardonnêche, declared the feasibility of cut-ting a tunnel through Mont Fréjus in the line where the Alps stood on their smallest base, between Piedmont and Savoy. In 1841 he actually published his convictions. The king, Charles Albert, of Sardinia, took up the question, and ap-pointed two engineers to investigate it.

Signors Maus and Sismonda, after a long consultation, decided in favour of the project. The former gentleman was a practical railway engineer, a Belgian, employed by the Sardinian Government ; the latter was a learned scientist, whose opinion upon the strata, etc., of the mountains was needed. Thus fortified, Signor Medail felt that success was at hand.

But success was still afar off.   A tunnel was possible *if,*—
and then came the objections.   Where can you find air to
ventilate the tunnel ?   And if you could find the air, how can
you sink shafts in the mountain ?   The line taken was correct,
but the idea, though feasible, was almost impossible of being
acted on, unless some means of cutting through the mountain
could be found.   To bore into it and mine it out would
occupy years and years, and probably result in failure.   The
question of shafting and working from these bases, in both

MONT CENIS, BARDONNÊCHE.

directions, was at once dismissed.   The mountain must be
pierced—how?

Yes, how many hands could be employed in cutting into a
mountain a railway-tunnel seven miles long ; and how were the
miners to be supplied with air as they advanced?   A curious
combination of circumstances solved the initial difficulty.

It appears that about that time (1855) an Englishman, Mr.
Bartlett by name, was greatly interested in an instrument which
he designed for perforating rock: a kind of steam-drill for coal-
mining, to supersede hand labour.   At the same time, three
Italian engineers, who had all been studying in England, were

deep in a project which they had conceived of using compressed air to drive a railway train up a mountain in the
Apennines.

Compressed air has considerable force when released. If
the air be subjected to the accumulated pressure of six atmospheres, it will, when released, perform work, if properly conducted. The air to be compressed must be forced into a small
space, and to effect this some power must be used. But
steam power means heat and fuel, and air to burn fuel. Thus
steam was put aside as costly, and water power substituted.

These young men, Sommelier, Grandis, and Grattoni, were
delighted when, at Genoa, they heard of Mr. Bartlett's invention. Thus compressed air could drive the perforating machine
in places where no air was available to keep a fire burning for
steam ! So they said,—

"Let us combine the ideas, and employ the drill with compressed air, produced by the power of water, plenty of which
is at hand ! "

The engineers submitted their plans, but without immediate
results. Troubles had fallen upon Italy, and it was not till
1857 that the Italians, under the ægis of Count Cavour, the
enlightened, began the surveys for the first great Alpine tunnel.
The Semmering Railway claimed precedence in 1854, with a
short summit-tunnel.

The surveys and the tracing of the direction of the line were
carried out under circumstances of some difficulty. The length
of the tunnel was estimated to be some seven and a half miles.
The district was wild and desolate ; the hamlets of Fourneaux
and Bardonnêche small and rude. To find the centre line by
which to guide themselves was the first care of the engineers,
and after much measuring they discovered that the tunnel must
be cut under Mont Vallon, in its centre.

To gentlemen who "live at home at ease" prospecting and
surveying have little danger or difficulty. A line can be taken
even over a mountain, and "there is nothing so very difficult
in that ! " Well, let any party go up and try to measure a

perfectly straight undeviating line over a mountain 6,000 feet high, exposed to storm and tempest, rain and wind, and snow; and then let them bring their calculations for the cuttings at two sides of a straight tunnel to meet exactly in the centre of the mountain, nearly 6,000 feet below the summit, and more than three miles from each end of it, in the bowels of the earth.

This was the little problem which confronted Signors Capello and Borelli, the engineers of the tunnel. Not only must the headings be driven from opposite sides to approach each other, but the cuttings must rise in equal proportions to the same elevation, and meet, more than a mile in perpendicular measurement under ground, at the same level. How well the surveys were made, and how correctly followed, may be gathered from the following extract from a newspaper of 26th December, 1870:—

"The working parties in the opposite headings of the Mont Cenis Tunnel are within hearing distance of each other. Greetings and hurrahs were exchanged through the dividing rock, for the first time, at a quarter-past four o'clock on Christmas afternoon."

The difference in the levels when the rock was actually perforated was found to be somewhat less than one inch, so correctly had the lines and elevation been traced and made. We will now briefly trace the manner and progress of this great Alpine tunnel.

The work was commenced in 1857, at the end of the proposed tunnel, by hand-cutting. At this spot, once a small and out-of-the-way village, houses and works sprang up almost as quickly as mushrooms. Canals for the conveyance of the water were commenced from the neighbouring stream. The headings were not commenced on the opposite side in Savoy until 1861, when the province was in possession of the French.

It was necessary to drive the tunnel so that the entrances should be as nearly at the same level, but with a slight gradient to each mouth from the centre. This small gradient was adopted to ensure the fair amount of drainage to each country. The line could have been made throughout on a gradient. The mountain base or level of the valley at Bardonnêche is higher than the Fourneaux side, and consequently the entrance to the tunnel on the latter is three hundred and forty feet higher than the level of the valley; a steeper grade of 0·022 to the mètre being made, against one of 0·0005 on the Italian side. One peculiarity of this tunnel must be noticed—the entrances were not eventually used for the railway lines. Short junction tunnels were constructed at the sides and the original openings are used only for ventilation. The railroad therefore does not run entirely in a straight line through the mountain, the curved junction tunnels being together 1,210 mètres in length, or 3,900 feet, the direct distance abandoned being 356 mètres on the north side and 261 mètres on the south side.

The mode of perforating the rock previous to the necessary blasting operations may be described, as it was subsequently the means by which all the other Alpine tunnels were executed, but by somewhat improved machines. The air was compressed at a distance of about half a mile from the tunnel. A number of tubes like organ pipes were connected like the capital U— in fact, siphons. In the base is a piston, which is moved backwards and forwards by a water-wheel. The legs are partially filled with water, and as the piston moves it drives the water up, compressing the air at the top of the water. The air when thus squeezed into about one-sixth part of its volume is admitted by a valve into a tank to be stored. The piston in its return movement performs the same operation in the other leg, and so on.

So the stream turned the wheel, the wheel acted on the piston, and forced the water up and compressed the air; thus the action of running water pierced the Mont Cenis Tunnel. The compressed air was conducted through a pipe on rollers, on supports of masonry, and with telescopic joints. The reason given for this arrangement was the rapid changes of tempera-ture, and when affected by heat or cold the portion of the pipe expanding or contracting was driven into the rolling portion by the slide joint, and pushed in or pulled out ; so the length and " play " of the long pipe was maintained.

THE FIRST TRAIN THROUGH THE TUNNEL, MONT CENIS.

The joints were almost completely air-tight, also, so little loss occurred ; and once within the mountain, the pipe—eventually more than three miles in length—was laid as usual.   The temperature in the tunnel did not vary much.

Those who visited the tunnel while it was in the course of construction will remember the covered drain in the centre of the footpath at the side. But later generations need to be informed that the culvert in the centre proved useful when any sudden fall of rock blocked the tunnel. On one occasion, at least, a gang of men was imprisoned thus, and ran considerable risk of dying of starvation till one of their number remembered the drain, which he and his mates then entered, and crawled out of in safety.

The perforating machine is a curious looking thing, possessing nine rods or drills, capable, each of them, of delivering two hundred strokes a minute, with as much force as, and with greater precision than, a man; the blow and the drilling are, by the machine, effected simultaneously.

In the somewhat confined spaces in which the machines worked compressed air was supplied, and the ventilation was complete; the air in the inner gallery being actually purer than that in the main tunnel. The perforators worked by the compressed air acting on pistons had no holidays; the men worked in "shifts" or gangs, but complete rest only occurred on Christmas Days and Easter Days for all those fourteen years.

The rock when punched into holes was considered ready to receive the powder. Each hole was blown clear by the machine first; and then, charging and blasting the centre series of holes, the miners proceeded to explode those surrounding the first space blasted, until a large portion of the honeycombed rock was displaced. The holes made in ordinary rock were some thirty inches deep. The displacement was therefore very great at each discharge.

Thus the Mont Cenis Tunnel was finished in time. The labour was great and continuous, and it must not be supposed that the machines lasted throughout. Those delicate tools were frequently replaced, and it was stated that one machine on the average "gave out" for every seven yards excavated. Nor is this surprising when we consider the action of the perforating drill, thumping and turning at the same moment,

delivering three blows every second, and revolving by the most
delicate mechanism after every blow, one-eighteenth of a revolution each time, every stroke being of an impulse of one
hundred and eighty pounds.

The railway was opened on the 17th of September, 1871,
amid great rejoicings.   The work had been carried through by
the Italians; for France, though willing to assist in the con-

RIGI ENGINE.

struction by contributing a share of the cost, opposed the line
in 1866, and afterwards, for fear that the Marseilles route to
India would suffer.   Her selfish policy succeeded in delaying
the enterprise, but not in stopping it, and was punished by the
initiation of the St. Gothard line, which placed Switzerland and
Italy in direct communication.

The works on each side of the Cenis Tunnel were carried out
in a similar manner; and now the traveller to Turin may pass
through the mountains with comfort, but not so pleasantly

perhaps as in the old days of the Fell Railway, which was constructed on the pass itself, upon Napoleon's Road.

Thus the first system of mountain railway was a success. To the uninitiated it appeared a daring work, but since its day the Rigi and Pilatus lines have snatched the palm of wonderment

VIEW ON THE RIGI RAILWAY.

and daring from the Fell Railway. Wherever we turn now-a-days the mountain railway confronts us, and we wonder what new modes of locomotion and progression are in the future.

The Mont Cenis Tunnel, which cost £3,000,000, was quickly followed by others. Notwithstanding the prophecy of a writer on the subject, who declared that the Mont Cenis Tunnel "as it is the first will be the last enterprise of its kind that will be undertaken for generations," the scheme proved to be only the first of a series of grand successes in engineering in the Alps.

After all, it is not wonderful that such interest should have been aroused in the Cenis Tunnel—an interest now generally abated. But it must be remembered that the pass is, as it was then even in a greater degree, an important one, that its dangers in winter and romances in summer were many. Some of us have seen the fatherless child carrying his little box of white mice come in, startled and miserable, to recount, in an almost understandable *patois*, how his father had been swept away by an early snow-slip and had perished on the pass.

The Fell Railway, too, zig-zagging up the road, from which majestic views are obtainable, possessed an interest of its own for travellers in search of the picturesque. The change from valley to mountain side, the vines gradually giving place to the chestnut and the hardier firs ; the wood then disappears ; the swiftly rushing stream, so business-like below, here in the upper slopes becomes a hasty, brawling, impetuous torrent, in a tremendous hurry to get down, leaping over the rocks and stones, and swirling round the more massive boulders amid the snows. All these impress us as we mount upwards.

Up in the solitudes we can, if we please, picture those who have passed by before the Fell Railway was in existence or the great tunnel was completed, and there were Alps no more. Putting Hannibal and his "vinegar" aside, we may find amusement in contemplating Horace Walpole, Evelyn and Lady Mary Wortley crossing the Alps here in *chaises à porteur* —"a low arm-chair," as Walpole describes it,—and his experiences were not the most pleasant.

This "low arm-chair slung on poles" was formerly the mode of transit across these lofty passes. "If I come to the bottom you shall hear of me," wrote Lady Mary to her friend; but she survived the dangers of the trip. In after years, about 1775–6, sledges or sleighs were used, for in the olden time travelling carriages were taken to pieces and packed on mules while the traveller was carried as described. Illustrious travellers were

fain to cross in those rough conveyances till Napoleon made
the road, over which he afterwards travelled with his Empress
Josephine to be crowned in conquered Italy ; but he was not
compelled to " slide down on his back," as he had been obliged
to do with his troops on the St. Bernard.

The diligence usurped the place of the mule, and the Fell
Railroad of the diligence. Firm in its tracks, of which there
were three, the little engine and train sped puffing up the
heights, " Excelsior ! " One foot in twelve was the rise, and it
appeared steeper. So winding upward we went upon the outer
fringe of the road, whence nothing to speak of intervened
between a slip and death. The experience was novel, but not
so greatly alarming as one we experienced on the zig-zag
descent of the Simplon to Brieg, when the drunken driver
of a diligence trotted down the mountain, the wheels within
a foot of the edge at times on the curves, as the well-trained
horses slung round the corners.

The curving, steep Fell Railway is no more, but the views
remain. The " Devil's Ladder " may still be ascended or
descended, but the railway has vanished.

AN ALPINE SCENE.

VIEW OF BERNE.

CHAPTER IV.

ST. GOTHARD, ITS DANGERS AND ROMANCE.—THE GREAT
TUNNEL, AND THE RAILWAY UNDERTAKING.

HE Mont Cenis Tunnel had not yet advanced to completion when the jealousy of the French Government urged Italy, Switzerland, and Germany to unite in a project by which an independent line of communication should be completed through the Alps. It was almost a matter of necessity, not only on political, but on commercial grounds, that a new line, not subject to French control, should be constructed between, say, Lucerne and Milan.

Such a stupendous undertaking as a railroad over and through the St. Gothard would never have been entertained if the Italian engineers had not already demonstrated its practicability. The idea was not new, but till Sommelier showed the way no one had deemed such a work possible. The Gothard mountains raise a bold barrier between the Swiss and Italian lakes ; a small rugged pathway first, then a road, then a splendid post-road, were necessarily made and used; but not many more than a hundred years ago no wheeled cart or carriage could traverse St. Gothard, and then the cost of transit was very great. But traffic was not suspended on that account. As in other parts of Europe, and in England, mules and pack animals wended their way over the rough places, skirting the torrents of the Reuss, and running many risks from avalanche, ice-fall, and landslips.

Records point to an annual passage of some sixteen thousand people, with perhaps half as many horses and mules, over this dangerous pass. From Fluelen by Altdorf up to Hospenthal, by the terrible Devil's Bridge, did the rider or pedestrian go in hourly peril of Nature or of mankind ; for the lonely places were haunted by robbers, and the echoes alone answered to the despairing cry for help. In the winter the passage was extremely dangerous, as it is still, even in the railway, at times.

The St. Gothard road was constructed in 1830; but excellent as it is, its dangers were, in the ante-railroad days, many and great. Let us take a rapid survey of the old path and road, which well reward our investigation, and those readers who will make the transit in October will have much of the excitement and little of the danger of the route. For about the middle of that month snow commences to fall, and then the " rutners," or roadmakers, in one sense highwaymen, turn out and commence their winter work.

The snow falls more heavily after a while, and the traveller will find himself placed, well wrapped up, in a sleigh. The road has, we will assume, been cleared by the rutners, who have packed the snow high on either hand of the track in places, in other places the edge has no protection, and the white mass overhangs the river in an immense cornice, which seems to be so firm, and is as unstable as water itself.

The sleigh runs merrily along with jingling bells. The beautifully dry, cold air tingles in your ruddy cheeks, and life

WINTER DILIGENCE TRAVELLING BEFORE THE TUNNEL WAS MADE.

seems most pleasant. The great hills, the dark stream, the rounded forms of the snow-clad rocks, the wondrous silence, all and each impress the voyager, who is struck with admiration and surprise. The top of the Pass is reached, refreshment is taken, and then down the sleigh rushes, zigzag perhaps, perhaps in a bee line, swinging dangerously, madly along. Eyes are closed, ears are deaf, the cold blast numbs the features, but a rubbing puts all right! Then you turn round, look back, wonder if it is all real! Have you actually descended so far in a few minutes? Is it yourself seated, tingling all over, with reflowing blood and with nervous apprehension? Yes; you are really alive and safe, and you make a mental resolution never to try the trip again. Yet you break the vow.

But there is a dark side to the St. Gothard also, when the gathering gloom and the slowly falling mist warn you, the traveller, to get on. You glance back, but the prospect is dull, indistinct, and bears a yellow tinge which is not agreeable. Your driver is anxious and silent; the horse struggles on, and looks nervous too. The silence is truly alarming, and a death-like stillness is spread around.

The clouds begin to descend and to shroud completely the already half-hidden peaks. The traveller wonders at the changing character of the landscape; there cannot be a storm because there is no wind. But, in a few moments, a sudden puff arouses the sleeping snow. It curls up in an eddy in front of the horse, which snorts uneasily, and pulls harder up the hill.

But nothing more happens—yet; the silence is again undisturbed. You go on. Yet again the sudden swirl of snow, right in your face this time! Get on, get on!

Then silence again for a while; but soon a low, gentle, but most weird and unearthly sound is borne to the ears—a kind of dirge of the snow-spirits, a "Keening of the Banshee!" The echoes of these voices appear to the now excited mind to be those of lost travellers, who are distantly beseeching help from you who will so soon need it for yourself. The cries increase, the wind adds a bass accompaniment, the clouds come lower, and then some icy flakes come darting through the air!

They sting horribly; but you must not turn. Go on, on as fast as possible, ere the track be hidden, the poles almost buried in the drift, and a mass of new snow envelop you! To halt would be fatal. The brain is already becoming bewildered;

x

drowsiness has begun, numbness is stealing over all the senses by the time the snow-clad horse, driver, occupant of the sleigh, and the vehicle itself come to a stop in a cloud of the snow at the Hospice, or refuge-house.

Then the dogs are on the alert; next day the face of Nature is changed once more. All is white, deep, snow, and

THE HOSPICE OF ST. BERNARD.

heavy masses threaten to fall. Then bells are muffled on the team; the driver and the travellers are silent; the loud-voiced lash no longer wakes the echoes of the Pass. The avalanche is hanging yonder, and a sound may bring it on the road as you creep like guilty cravens round the bend and down the hardened snow-track.

Under such circumstances, travellers used to cross, and still may cross, the Alpine passes where the railroad now is in evidence. The days of Felix Lombardi were those of adventure and romance. The St. Gothard Hospice was under his charge for nearly three-and-twenty years, and he was a man of might in his way—a host in himself. From humble beginnings, he became a sleigh-driver, then took to tailoring; then, when the post road over St. Gothard was made, he drove a diligence; brewed beer in Bellinzona, and became the government inspector of traffic in the Pass. Many a gruesome tale could Lombardi tell of the road over which he diligently watched so long, of the travellers he had rescued, of the escapes, and also of the fatalities in his experience, and of history. He enlarged the hospice, and was as good a host as any one could desire. His care of the half-frozen travellers in those earlier days of travel was beyond all praise.

In later years the possibilities of a railroad across the Pass began to be discussed. The Brenner, the Semmering, and, later still, the Mont Cenis had been conquered; why not the St. Gothard? Those were disturbed days, and not till 1869, when a conference was held at Berne, was the line approved. In June, 1870, the lines were agreed upon; Germany, victorious, was resolved. The line was decreed, but not actually commenced actively until the autumn of 1872.

The railroad was projected from Lucerne to Lugano, and thence into Italy, which with Switzerland and Germany made mutual arrangements.

The treaty was signed in due course, and the estimated cost of the tunnel was put at one hundred and eighty-seven millions of francs; but this sum was found to be utterly inadequate, and at least another hundred millions of francs was deemed necessary. Naturally this announcement filled many hundreds of shareholders with dismay; a company had been formed, and the prospects of the stockholders became clouded. At the time France and Germany were flying at each other's throats, but, notwithstanding all drawbacks, Switzerland and Italy did their best ; the former state, particularly, made great efforts. M. Favre, of Geneva, had undertaken the contract, and meant to complete it by the 1st October, 1880, the date named.

The experience already gained in the construction of the Mount Cenis Tunnel was utilised to the full. We need not detail the method of working, which was by drilling, etc. The compressed air for the boring tools was conveyed into the tunnel in an iron tube, and dynamite was largely used in

ENTRANCE TO THE GREAT TUNNEL, ST. GOTHARD.

blasting. The materials to be used in the tunnel and those which had to be carried out of it were transported by an air locomotive, an ingenious invention to obviate the smoke and heat necessarily evolved in the employment of steam. For these engines a special supply of air, compressed to the power of 10–12 atmospheres, was used.

These boilers were not picturesque, but they performed their work well. The dimensions of the tunnel are much the same as that of Cenis—twenty-six feet wide and nineteen feet high. The men worked for small wages; and though the average payment of the miners was about three and a half francs—say three shillings a day in British currency—they boarded themselves, and managed to exist and to work on half that sum.

For this wage they encountered many dangers, and had some accidents as well as several escapes. But the most curious scene connected with the work was the strike amongst the men, which was thus described by one of the foreman :—

"I was standing superintending the work, when suddenly there arose a tremendous din and shouting. Turning in the direction of the noise, I could perceive in the tunnel hundreds of twinkling lights approaching rapidly. Fully a thousand men, were rushing out, and I could distinguish cries of 'Gas!' 'Save yourself!' 'Run for your lives! Quick!'

"You could hardly have heard this before the men came rushing out, half-naked, shouting, and jostling each other. Some held kerchiefs to their mouths, some their hands, but all were surging and pushing out without any pause, and without giving any definite clue to the cause of the tumult.

"Of course, I ran too. Carried away by the mass, and full of apprehension, I ran as fast as possible after the crowd, and fancied all kinds of evils were coming upon me. But the atmosphere of the tunnel told severely. It was impossible to run far so fast; and in a short time, thoroughly exhausted, I sat down panting, to await death, for I could go no farther then.

"The men had gone. There was I alone, seated upon a boulder, almost breathless, the foul gas approaching with certain, slow advance, and I prepared to die, if I must. I gasped, and fancied myself suffocating in the blackness of the tunnel; bethought me of friends and relations, my wife, my young children ; what would they do when my dead body was found next day—or, perhaps, not for many days—perhaps never?

"It was horrible, terrible; such a ten minutes surely few men

have survived ! But the gas did not come—suffocation had
not begun.   Could I not rally and make another effort to
escape ?   Yes ! I would try.   Up, up, run on, I said to myself.   Race for your life, and for your wife.

"I did.   I ran on, and, to my delight, saw the entrance becoming larger and larger ; fresh air met me at last.   Plunging
on, I gasped at it, and felt reanimated.   Behind me extended
the awful blackness, the very shadow of death.   In front,

A GALLERY, ST. GOTHARD'S RAILWAY, AFFORDING PROTECTION
FROM AVALANCHES.

sunshine, the fresh mountain air—life !   How I welcomed the
hill side, the ugly buildings, the noisy miners who surrounded
them,   still clamouring, and at times threatening.   The men
extended themselves across the ways, and clearly some disturbance was afoot.

" 'What is the matter ?' I inquired.   ' Is any one lost in the
explosion ? '

" ' Explosion !   There is no such thing.'

" ' No explosion ? No gas—no injury ?'

" ' Certainly not, sir.'

" ' Then why did the cowards run out so fast and give the alarm ?' I asked, forgetting my own headlong race and terror.

" ' Because they wanted to clear out of the works. They are on strike now, and demanding terms. They refuse to return, and will have their rights.'

" ' Then "the gas" was a false alarm?' My fancies had excited me. I was angry with the men, and wished them punished. But my feelings changed when the soldiers of the Canton arrived and fired upon the mob, killing several, and wounding others. These were harsh measures, but seemed necessary— though I hoped the soldiers would refrain. The miners returned to the work after a while ; and that was the only strike of workmen in my time, or in any other, I believe."

Dynamite, as stated above, was the chief material used in blasting, but on one occasion it was used illegitimately by a revengeful workman, who cast a cartridge into the rooms generally occupied by the overseers. Providentially the officers happened to be out at the time, and when the smoke had cleared off the huts were in a condition suggestive of match-wood.

Some accidents also happened. On one occasion some cartridges exploded, from some undiscovered cause, and blew the gang of men within the influence of the dynamite literally to fragments. Nought of them, save some feet encased in miners' boots, was ever found. Nevertheless, the tunnel progressed, the headings and borings were pushed on at each side, and at length the miners came within a measureable distance of each other.

On February 29th, 1880, the workmen, who had been labouring so diligently, met. The thin strata of rocks which had separated them for the last day or two previously was then pierced. The men cheered and gripped each other's toil-stained hands through the aperture. But no one attempted to pass ; even welcomes were quietly stayed until a portrait of M.

Louis Favre, the late contractor for the work, had been passed to and fro between the gangs.

Why did not M. Favre himself appear to witness the triumph of his skill and anxious labour? you will ask.   Alas! he was dead.   Riot, strike, the burning of Ariolo village, and their worries and disappointments, his own marvellous exertions and numerous troubles, had told upon him.   Before the tunnel was completed he had succumbed.   Visiting the works one mid-summer day (19th July, 1879), he was reckoning upon the completion of his grand work.   But even while in the tunnel he staggered and fell into the arms of his companions.   They bore him out, but the doctor pronounced his seizure apoplectic.   Every remedy was tried on the spot; ready hands and kindly faces ministered to him, and surrounded his rough resting-place.   But he died very soon after his seizure on the scene of his grand undertaking—an undertaking of which he has been forbidden to see the end.   .

Like another Moses, he had led his people from conquest to conquest; had triumphed over many foes, and quelled many threatened dangers and mutiny amongst them.   Still, he was only permitted a glimpse of the promised land, so to speak. He had brought his task almost to a termination, yet he was not destined to see it completed.

But his remains were honoured.   The Canton decreed him a public funeral; the men paid him a most eloquent tribute, and he went to his grave sincerely, deeply, regretted.   This is the reason why only a portrait of M. Favre was first handed through the aperture, and why no man passed until he, in his portrait, had preceded him.

So some miles of solid ground had been tunnelled, and so accurate had been the surveys and the levels—only a difference of two inches was shown in the level, and only thirteen inches in the direction of the headings, in the whole nine miles of tunnel, which proved to be actually only twenty-five feet shorter than the length originally estimated.

Although the tunnel was pierced, the line was not ready, nor was the tunnel itself in a condition for the rails to be laid in it. Not until January 1st, 1882, was it ready for traffic; but no one will suppose that it is the only tunnel. We hear of the St. Gothard Tunnel most, but in the length of the mountain railway are fifty-six in all, some of considerable length. Of these the traveller may reckon twenty-eight on the Italian

VIADUCT NEAR WASEN, ST. GOTHARD ROUTE.

slope, and twenty-seven on the Swiss slope, the summit tunnel (St. Gothard) being 16,308½ yards long, and in the centre it stands 3,787 feet above the sea level.

To the tourist the scenery, and the very peculiar method by which the railroad doubles upon itself, like an enormous snake, will appeal. The corkscrew, or "helicoidal," tunnels are most marvellously constructed, ascending in the interior of the mountains. Occasionally the train crawls out; a glimpse of a lovely village, a brawling torrent, or a church tower is gained. Into the dark again, and on again emerging the same village is seen again in a different aspect from a greater elevation; and it needs all our toleration to credit our friend's assertion that the train doubling apparently after us is, in reality, one which we passed in the last tunnel on its way down!

Since the opening of the line many improvements have been made. There is a double line throughout the tunnel; there are lamps at frequent intervals, and fresh air is abundantly supplied. The traveller can stand in the gallery of the carriage and look down from outside the train upon the grandeur of Nature, and compare man's greatest efforts in engineering with work done by the fingers of ice, and snow, and tempest, above and around him.

We must here conclude our unofficial account with a few facts. The St. Gothard Tunnel, you may remember, was commenced on 4th June, 1872, at Göschenen, and on the 2nd of July at Airolo. The cost was £2,320,000. The cost of the entire railway was £9,520,000, or say two hundred and thirty-eight millions of francs. The greatest gradient is one in four, and the number of miles occupied by tunnels is twenty-five, besides nearly fifty bridges and viaducts.

Four thousand men were employed. The line occupied nearly ten years in construction, and was first worked on the 1st of June, 1882. The steepness of the ascent is, on the average, six per thousand feet on the Swiss, and two in a thousand on the Italian side of the tunnel, which is 2,930 yards longer than the Mont Cenis; and the time occupied in transit is, in fast trains, about fifteen minutes.

To readers who desire a most pleasant and picturesque ride we say, go over the St. Gothard by road or by rail; either will prove delightful in fine weather.

THE THAMES TUNNEL, 1843-65.

CHAPTER V.

FORMER TASTE FOR TUNNELS.—THE SEVERN TUNNEL AND ITS
ROMANTIC SIDE.—ACCIDENT AND ADVENTURE.

ITHERTO we have dealt only with the most important tunnels through rock or mountain inland
—"underground" merely. In the next portion of
this section we propose to consider the many
incidents and curious experiences connected with some of the
subaqueous tunnels—those under rivers and channels of water.

The foremost of these we need not enlarge upon. The
Thames Tunnel, the wonder of its day—if a tunnel can be said
to have any day—has been described so often, and is so well
known to the present generation, that we need only refer to it.
Brunel drove it under the ægis of his "shield" at a great cost
—about thirteen hundred pounds per yard. It was originally
a footway ; the East London line now uses it. The Hudson
Tunnel and the Mersey Tunnel offer other examples, as we shall

see ; but the Severn Tunnel is the most important of all, and
we will commence with it presently.

In the old days of railroads, it will be remembered, there was
on the part of a large majority of the public a great objection
to the tunnel. People appeared to have a rooted dislike of
such underground passages, and hardly considered the resem-
blance which existed between travelling by night and travelling
for a short time in a tunnel.

The changes of temperature, the "thunder peals," the
"sudden plunge into almost total darkness," the rattling
wheels, the roaring rocks, clanking chains and dismal glare,
were some of the epithets bestowed upon the passage of a
railroad tunnel ; terms which would almost have been appli-
cable to Pandemonium itself.

Nevertheless, there were advocates of tunnels who described
them as "pleasant," dry and agreeable ; free from smell, and,
virtually, rather enviable places on the whole. These people in
their turn ridiculed the danger, and scoffed at the fears of the
public. One amateur advocate of subterranean passages was
so excited by the idea of the tunnel that he endeavoured to
outdo Stephenson. This amateur resided near Liverpool, and
was fired with emulation of the Lime Street Tunnel. He
possessed somewhat extensive grounds, in which he very
quietly and privately proceeded to dig and construct a tunnel.

No one had any idea, save his own assistants, that he was
thus employed so near the railroad tunnel which Stephenson
had constructed. But one day the great engineer, when
inspecting his handiwork, was surprised to hear his name
called. He gazed round him in astonishment. His companions had also heard this mysterious voice, and were not
free from alarm.

"How are you ?" repeated the voice. The men looked up
in the direction of the sound, and perceived a face peering
through a hole in the tunnel wall. This was the amateur ; but
he was soon compelled to close his shaft and bore in another
direction.

We will now proceed to consider some of the most interesting specimens of the Tunnel family in England.

THE SEVERN TUNNEL.

As we have seen in the foregoing pages, the labour of tunnelling had been greatly diminished by modern appliances. In the case of the Kilsby Tunnel, one of the most remarkable of our underground ways in early days, the outbreak of a spring of water damped the enthusiasm, and nearly swept away the hopes, of the directors of the London and Birmingham Railroad. The directors of the Great Western Railway, fifty years later, when engineering science had reached almost its highest point in this generation, were destined to suffer a similar experience.

The traveller who now travels from Stapleton Road (Bristol) to the Severn Tunnel Junction, on the opposite side of the river, is carried smoothly by the Great Western Railway down a long and somewhat steep incline through a cutting, until a warning whistle from the locomotive announces, almost simultaneously with the sudden plunge into darkness, that the train is in the Severn Tunnel, under ground and under water ; and in the tunnel it continues its way for more than four miles, when the engine emerges again into daylight, and labours up the incline on its way to South Wales.

That is what the present-day traveller experiences. An easy road, an easy pace, no hurry whatever, a neat-looking incline, a well-proportioned tunnel, and he, the said traveller, perhaps never bestows a thought upon the years and years of patient thought and labour which has resulted in the present quick passage from shore to shore beneath the bed of old Severn. Such are the results of latter-day engineering.

Before this splendid work was finished passengers for South Wales were conveyed by the Old, and subsequently by the New, Passage routes. The Severn being nearly two miles wide at this passage, it was no small task to cross under unfavourable circumstances. The Great Western Railway, led by Brunel,

went round by Gloucester city, and thence along the course of
the river—as it still goes —to Chepstow and Cardiff. The cost of carriage for merchandise and minerals some forty miles to Gloucester was out of the question, and small steam boats were used to transport passengers and cattle or produce across the ferry or passage by which the Bristol and South Wales Railroad Company conveyed their customers to Portskewett by means of G.W.R.

Portskewett, or Portiscwit—"the port under the wood"—is a small place, remarkable for the Roman encampment of Sud, or South, Brook on the cliff. This encampment served to defend the port under the wood, and in the twelfth century A.D. one John Southbrooke had "a grant of key-bote and house-bote" for his mansion there. Hence the name of the small so-called village, destined to grow and multiply into a most important place in after years.

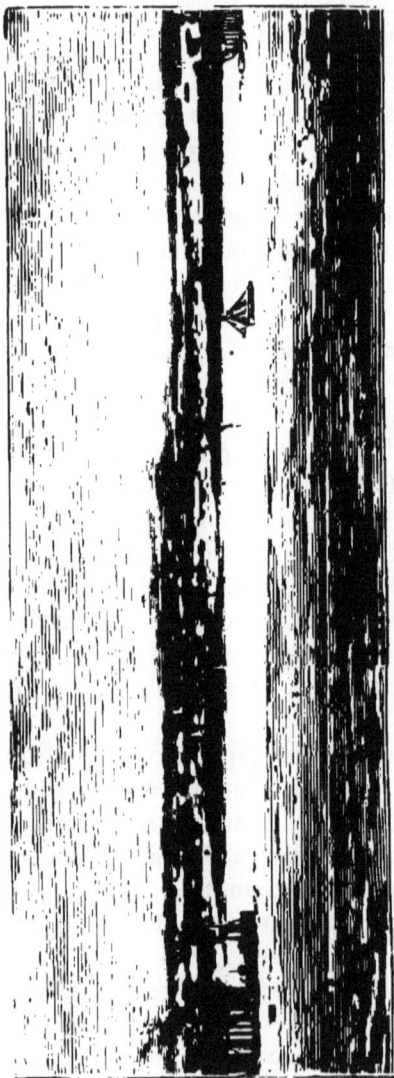

THE STEAM FERRY TO PORTSKEWETT.

The village of Sudbroke not very many years ago consisted of a name, a ruined picturesque chapel, which still stands, and a farmhouse or two with outbuildings. From this spot to the farther shore the Severn Tunnel was driven to meet the workers in the opposite direction.

But though the idea of connecting the Great Western with the South Wales lines was conceived by Mr. Richardson thirty years ago, time passed and the steam ferry held its own—a very precarious hold at times. In bad weather it was far from pleasant to embark at low water after quitting the train and standing waiting, perhaps in rain or snow, until the baggage had been transferred to the already crowded and not always savoury steamer. Sometimes the rushing tide or opposing wind, or both together, made the transit more lively than was desirable, and the ascent of many steps after such an experience, short as it was, was trying. People grumbled, submitted, or went round by Gloucester, but yet the tunnel scheme came no nearer a practical solution until 1871, when Mr. Richardson, of the Great Western Railroad, lodged the plans for his project. There was then no bridge over the river nearer than Gloucester. Traffic had been carried on by the various " Passages " at the Lodes at Newnham, at Purton, at Aust and New Passage. These, as many will remember, were not devoid of danger, for high tides and the tidal waves are not to be despised on the Severn. The contest of fresh and salt water, termed the " Eager," or " Eau-guerre," made the river very rough. So until Mr. Richardson brought forward his scheme communication was tedious and tiresome 'twixt England and South Wales—and is not rapid now. No one can praise the Great Western, once the pioneer of speedy travel, for their service to South Wales in this year 1892.[1]

The plans were deposited in Parliament, the Bill passed in 1872; and the work was commenced in 1873, when the tunnel was begun in an easterly direction from the Monmouthshire bank.

[1] A service of fast trains is wanted ; but " there is no competition," say the people, " and the G. W.R. won't move."

The initiation of the works for the great tunnel completely changed the aspect of the place. The village, of a church and farm and a few cottages, found itself famous at one stroke of the engineer's pick. Sudbrook and its inn of "Black Rock" became somewhat more than a name; new cottages sprang up, navvies and gangers appeared, engineers and their foremen dropped down into the space, a tramway was laid "in no time," and almost before the villagers had time to rub their eyes and enquire if there "was visions about," engines, pumps, and building material, mining tools and machinery came over the road.

But, unfortunately, the speed was not maintained. The directors of the *Flying Dutchman* made little progress with the works; whether because of the cost or not does not appear. Their staff did not perform miracles, and after a leisurely boring of some four and a half years' duration, they decided to let the contract. This was quite characteristic of the magnificent and haughty line which hurries not itself for anyone, in anything.

Festina lente is an excellent motto—but it is not suited to a railroad company.

But even the scope of the contract was itself contracted. Shafts were sunk by the small "tenderers," while the company, having declined a highly estimated contract, continued to drive the heading themselves for another period until a serious irruption of water flooded the workings, from which the men narrowly escaped with their lives.

The consulting engineer, Sir John Hawkshaw was summoned to the rescue. He advised having a contract made, and in accordance with his advice, Mr. T. A. Walker—who, alas! is dead—was consulted as to his former tender, which had been declined. Arrangements were made; after a while the contract was signed, and Mr. Walker began the herculean task which will be connected for ever with his name; and the incessant labour that it imposed upon his most conscientious mind, probably, hastened his end.

Mr. Walker lost no time. He had engine houses built for

the pumps, and made his plans for checking the springs. The immense exertions made to clear the tunnel cannot be recapitulated here. A pump capable of raising three hundred and seventy gallons of water at one stroke was employed, with other smaller ones, and then the door in the heading had to be shut, so as to dam the water back.

But who could or would close the door? The difficulties

CUTTING ENTRANCE TO SEVERN TUNNEL.

were enormous. The divers were plucky and willing, but it required more than ordinary courage in a man to proceed three or four hundred feet from the bottom of the shaft, under thirty feet of water, draw a length of air-hose behind him, close the door, and by other means completely cut off the communication with the heading beneath the Severn.

To bring the matter to a safe conclusion three divers were employed, two to drag the hose along and pass it to the leader, who proceeded, of course without any light, in the direction of the open door which he had to shut and fasten, as well as to arrange the sluice-valves, etc., and to sever the connection entirely. The leader, Lambert, tried, but was unable to drag the air-hose after him beyond a certain point. He sat down and tried again after this rest under water surrounded by all

THE SEVERN TUNNEL DIVER IN GREAT ENGINE PIT, PREPARING TO DESCEND TO RECTIFY THE MACHINERY.

the rubbish, broken-down " skeps," or trollies, tools, etc., which one expects to find in an uncompleted tunnel.

He failed—no wonder ! But after a while Mr. Fleuss was sent for, and his diving apparatus was tried. This arrangement contains compressed air in a kind of knapsack; and this air can be supplied by the diver to himself as required, there being purifying chemical substances in the helmet to counteract the vitiated respirations, and thus the wearer of the dress can continue to breathe the same air without injury. Mr. Fleuss tried, but failed, and then Lambert donned the dress and started by himself into the passage.

The intrepidity of this man cannot be over praised. In a novel costume he went alone into the depths, amid all kinds of obstacles, depending for his life upon the small pipe which conducted the air to his helmet. Should this pipe break, death in a terrible form would immediately overtake the diver. But undeterred he proceeded, and though he did not succeed altogether in his task at the first attempt, he completed his laborious business on the second occasion, when he remained under water nearly a hour and a half.

This was the beginning of success. The water was reduced, the great spring was conquered, the streams and wells returned to duty ashore, for they had struck during the release of the Great Spring, which tapped them all; and the year 1880 ended more cheerfully than at one time had been expected.

By this time the workers were in full swing. Schools, mission-house entertainments, and such care for the miners, their wives and children had been taken, that a small town had arisen. A coffee and smoking-room were there, a club-room and neat stone or brick cottages. A hospital was there also, with the necessary plant on both sides of the river for electric lighting, pumping, and excavating work, which, with the exception of the pumping, was intermitted on Sundays, when services were held, and Sunday-schools attended.

Let us go in imagination down the shafts and peep into the workings as they were in the course of the construction of the tunnel. The shaft was not deep, and we found no difficulty in the descent; but it was very necessary to be clothed in suitable overcoats or oilskins, known as "donkeys," and slouch hats, to protect our dresses and our heads. All visitors who had any respect of person put on these disguises, which were almost impenetrable.

The descent was easy and rapid, and when we reached

the tunnel-level of the tramroad, we perceived a vaulted area, well lighted by electricity; while beyond, high up, twinkled

PUMP FOR CLEARING TUNNEL OF WATER.

many stars. The queer shadows, the gleaming water-current, the strong light, and the distant hum of men at work underground were somewhat puzzling to a stranger, and tended to arouse a feeling of awe—a weird dread of the unknown.

Nor were such feelings unreasonable. We, who now travel leisurely in well-lighted and padded carriages, under the Severn's rocky bed, do not realise the sensations which the visitor to the tunnel in construction experienced. There were the trollies or skeps drawn by ponies who never enjoyed daylight, save on Sundays; waterfalls plashed, and streams dashed out of the very rocks, regulated by the engineer "as required."

Overhead, on scaffolding, were men at work cutting the rock, and one could not help thinking, "suppose they cut too far, and let in the sea!" But it was not salt water but the fresh which the workers had to contend against.

The noise caused by the blasting was very alarming, the echoes continuing to roll in thunder-peals for some time; the concussion must have been very great on such occasions, and perhaps to this heavy firing is to be attributed the breaking out of the springs in the "faults" of the rock, even as heavy rain clouds are depleted by a royal salute, or a sham fight, or an explosion in a dynamite, rain-making, balloon.

However that may have been, on the 11th October, 1883, the progress which had been made and which had continued, despite of strike, panic, and accidents, was checked by the old enemy, water. The year 1882 had been a troublous one; but notwithstanding perils by water and by fire, by ill-feeling of men, and unfounded alarms, the year 1883 was an excellent one until, on the evening of the 10th October, the men at work on the Monmouthshire approach remarked upon the inroad of a stream of water which was percolating through the blasted portion of the rock a little to the westward of the place where the Great Spring had formerly broken out.

For a few moments the miners did not "fash" themselves about this, but ere fifty seconds had elapsed the rock fell out, the water rushed in, and a torrent of bright, pure water rolled in "like a great horse" as the men expressed it. So great was the force that the men and the skeps were carried bodily out of the section into the finished tunnel through the opening, and there in shallower, because extended, stream the men, bruised and half-drowned, managed to pick themselves up again, and made gallant efforts to shut the door which led into the heading.

But in this they failed. The water continued to gain rapidly

SCENE AT THE PUMPING SHAFT IN FIVE MILE FOUR PIT, MONMOUTH-
SHIRE BANK.

upon the pumps, and fell into the lower levels in a cascade. An increase of four feet an hour, and 27,000 gallons a minute, represents a considerable body of water, but it had to be grappled with. A wall had to be built under the middle of the river to check the advance of the water which had found its level in the tunnel, a mile and a half of which was flooded.

The pumps continued to work, and though one broke and the Sudbrook works were "drowned," the work went on in other places. No one after a while appeared to feel any alarm; there was anxiety as to the depth of water, but no serious consequences of any kind were really apprehended. Certain rumours regarding an expected high tide did not meet with much attention. What harm could it do? No tide had ever yet succeeded in reaching the Marsh-pit shaft in which work was being carried on, and so the men went down as usual to the night shift duty at seven o'clock. A considerable stretch of the tunnel had been completed there. The men quitted their homes—cottages which had been erected on the flat ground by the river, between it and the shaft.

With evening the south-westerly wind had increased. Great gusts, and longer blasts of sustained force, came sweeping up the estuary; the clouds, banked up with impending rain, flew faster from the sea and hurried inland to escape the tide which, like a wall, rushed in between the narrowing banks, and came up the Severn.

The Severn tides are reputed the highest in Europe. The "bore," or tidal wave, rivals, if it does not exceed in volume, that of the Seine, and the level of high water at Chepstow is at times fifty feet above the normal level of the stream. Urged then by a boisterous wind and equinoctial influences, the tide came on, and a terribly wild scene ensued.

A liquid wall came roaring up the river several feet high. It was late in the October evening, darkness had set in, and added to the horror of the scene when the wave arrived. The warnings were not unnoticed, but no one believed that the brown and muddy flood would devastate the tunnel. It was not unexpected when the ferry steamer, broadside on to the tide and wind, was carried, helplessly splashing, some miles up stream; when the pier, which she failed to "make," was undermined, and the platform and its rail swept off in triumph to Gloucester.

It was nothing to the miners, at first, that the banks were

overflowed below, that cattle in the fields were caught by the flood-arms and swept in a death embrace beneath the salt water. But when the watchman saw the roaring bore approaching foam-capped; when he perceived in the gloaming the ghostly phosphorescence which marked its approach; when, as he gazed, the wharf at Caldicot disappeared, and nought but Severn could he see; then he gave the alarm with all the fervour of fear, and "the Eager!"—the cry of danger—was raised.

Too late! The "solid wall of water" as it has been curiously termed, came on, and in a few minutes had invaded the flat marshland, had poured into the miners' houses, and risen above the beds in which the young children were peacefully sleeping. Frantic mothers rushed to pluck them from the cots, and the astonished babes, dripping in their scanty clothing, were promptly placed on shelves and dressers out of the reach of the invading sea.

The rush was tremendous. Five feet of "solid" water is no respecter of furniture, of goods and chattels, nor even of human beings. The chairs broke their backs, tables' legs were smashed in their watery struggles, and not content with this wholesale destruction and alarm in the houses, the tidal wave sought the bread-winners in the pit down the Marsh shaft, some hundreds of yards away. There was no time to inform the miners, eighty of whom were in the Tunnel works. The first intimation they had of the occurrence was the noise of the contest between fire and water. The boilers and furnaces of the pumping and winding engines resented the influx, and hissed loudly at the enemy; but the force was too strong; the fires were promptly overcome, and the victorious tidal-wave swept on into the pit, falling down in a torrent headlong, full fifteen fathoms. The Great Western line also had by this time disappeared under the flow of water; the tram-roads were no more seen at that hour; the cuttings were filled up; and trucks stood in the darkness, surrounded by a tempestuous sea, where they but ill bore the morning's reflections in the lake-like expanse in which they were left by the tide.

CONSTRUCTION OF THE TUNNEL ON THE GLOUCESTERSHIRE SIDE,
SHOWING THE SYSTEM OF TIMBERING.

But, ere this, the state of affairs in the shaft and in the tunnel
had become very serious. The miners, surprised by the sudden
cascade of water, made violent efforts to escape up the ladder
to the surface. A few succeeded ; one certainly was torn from
the steps by the cascade and killed. The great majority of
the men retreated before the flood, up the incline of the
finished portion to the unfinished upper end of the heading,
where, damp and terrified, they awaited their fate. The
electric lamps, in addition to their candles, showed them the
steady advance of the water, which rose by degrees until it
extinguished the lamps, and left the imprisoned miners in
almost total darkness, the candles being their only light in the
immense vault.

There they remained. One man was with difficulty persuaded not to swim to the shaft ; he said he would sink or
swim, but his companions dissuaded him. He did eventually
reach the ladder, and afterwards the " bank," but the alarm
ere this had been, though with difficulty, carried to Sudbrook,
where preparations for rescue were made. The receding tide
had made approach easier, but there still remained a great lake
three feet deep.

Volunteers were called for ; miners came hurrying up, and
made their way to the mouth of the shaft, but none attempted
to descend until one of the officials announced his intention to
go down to the rescue of the men in the pit. Some of the
miners then agreed to accompany their foreman, and his
example fired others of the staff. A party descended into the
pit, down which the water was still pouring apace, nearly
drowning the rescuers.

The flood had reached to within eight feet of the crown of
the tunnel arch, and there were no means of communicating
with the men imprisoned. A raft was impossible, a boat
almost out of the question. So the would-be rescuers stood in
the cascade on the ladder absolutely helpless, but not hopeless.
They reascended, sent to the " Pill " for a punt, and after considerable trouble, let it down lengthwise to the water.

As this was being done, the rope broke. A shout from the unfortunate men, still on the ladder in the narrow shaft, proclaimed their fears of death. Had the punt not turned in its fall, and wedged itself firmly across the opening, everyone on the ladder must have been swept off it into the water and drowned.

Most providentially this further catastrophe was averted, but considerable delay was caused. Meanwhile, the miserable miners below were beginning to despair of rescue, for they were of course ignorant of the cause of the delay. After some time the boat was righted, let down, and launched in the tunnel, and the rescuers, with lighted torches, went up to release their companions.

This was no easy task. To propel a boat amid a mass of débris of wood, floating trucks, timbers, and logs, was no child's play. Three men only—the fireman, and two others—went in the boat; the rest of the rescuers made a staging at the bottom of the ladder, whereon the rescued could be received and assisted to the top.

The brave pioneers advanced into the tunnel, but soon found a barrier opposed to them. In the distance they could perceive the twinkling lights, and could hear the voices of the imprisoned miners, but could not reach them until the beam that stretched across the tunnel-arch had been cut.

The boatmen returned to the shaft, sent for a saw, and again proceeded. After some hard work the beam was cut under water, as the workers thought—but no ! The timber defied their efforts. Again they tried, till exhausted they could work no more in such a cramped position. There was only one resource left ; the beam had been nearly severed. Could not they push the boat through the remaining splinters, which the saw could not reach ? They would try. A strong push and a pull together ! The wood yielded, the bluff punt crashed bravely through, and in a few minutes the three men were assisting their comrades from their rocky perch.

Even then the danger was not over. The boat could not get

close to the resting-place, and a rope had to be requisitioned.
By its means the men were eventually taken off, and brought
up the shaft to the surface in safety, about five o'clock on the

SIR DANIEL GOOCH.
From a Photograph by Messrs. Hill & Sanders, Eton.

morning of the 18th, having been imprisoned ten hours in the
tunnel, wet and miserable.

Then the pumping process recommenced, the pit was drained,
and strong measures were taken to conquer the Great Spring,
which a second time asserted itself so inopportunely.   The

spring was shut in by walls, and the rest of the tunnel was proceeded with. When this was finished, the spring was taken in
hand, firmly and finally shut out entirely from the works, the
underground stream being pumped into the Severn.

On the 5th September, 1885, a party, with the late Sir D.
Gooch, proceeded through the tunnel in a train. Now, the

CHANNEL TO CARRY WATER PUMPED FROM TUNNEL.

trains to and from South Wales traverse it daily, as a matter of
course, and few passengers bestow more than a passing remark
or a passing glance at one of the greatest achievements of
engineering in the world.

The actual length of the tunnel under the Severn is 7,664
yards—a semi-circular arch from 27 to 36 inches in thickness,
and of 26 feet diameter. It is lined with glazed (vitrified)

brick, and ventilated by a Guibal fan, by means of ventilating shafts, one at each end of the tunnel, and there are also pumping shafts to drain it. The top of the tunnel is just fifty feet under the bed of the Severn, and one hundred and forty-five feet below high (spring) tides. There are some bridges under and beneath the railway approaches to the tunnel, in the construction of which, and in the tunnel itself, more than seventy-six millions of bricks were used. Some three thousand men were employed, the largest day's pay amounted to £4,372 13*s.* 9*d.*, and the average daily quantity of water pumped from the Great Spring was twenty-four millions of gallons.

The first passenger train went through the Severn Tunnel on the 1st July, 1887 ; and as there are interior block signals, the fears of accident are reduced to a minimum. From our own experience, we can pronounce the passage of the tunnel easy and no more unpleasant than the transit on the open railway.

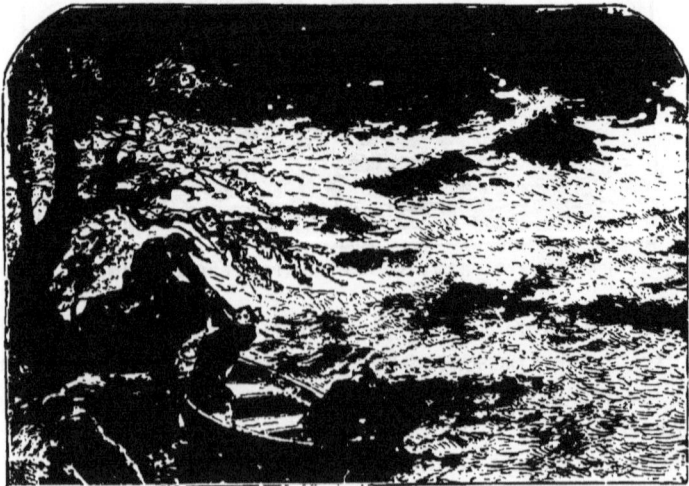

A TIDAL WAVE ON THE SEVERN.

INTERIOR OF BOX TUNNEL, NEAR BATH.
(*Broad Gauge.*)

CHAPTER VI.

OBSTINATE NORTHAMPTON.—THE KILSBY TUNNEL.—A TERRIBLE
ENEMY.—DESPAIR AND DETERMINATION.—TERROR AND
TRIUMPH.

HERE are few more remarkable instances of Railway land-tunnelling in England than that of Kilsby, the famous subterranean passage constructed by Robert Stephenson when he was engineer of the London and Birmingham Railway. As many people are already aware, it is to the prejudices of the Northampton worthies that we are indebted for the tunnel, which cost the Railway Company such an enormous sum. The inhabitants of Northampton stupidly declined to permit the line to come near their town, and so the hill near Rugby had to be tunnelled.

There was no alternative. Northampton remained obstinate. It didn't want a line, not it! But in after years it changed its foolish mind, and begged the North Western to help it. Robert Stephenson repulsed, had to proceed direct, and he projected the great Kilsby Tunnel, 2,400 yards in length.

It was a difficult undertaking, and the engineer could not say that he had not warning. The celebrated Dr. Arnold, of Rugby, said : "I understand that you intend to carry your line through these hills. I shall be much surprised if they do not give you some trouble."

Stephenson faced it. He took every precaution to ascertain the quality and nature of the soil and strata. Trial shafts were sunk, and the information was gained. The contract was let for £99,000, and the work began. Many—no less than eighteen—shafts were made by which the material bored and cut away could be brought out. So the work began.

But it had not proceeded very far when a quicksand was reached, and further investigation showed that it extended for some four hundred yards between two of the experimental shafts sunk to ascertain the construction of the hill. Such a discovery meant ruin to the contractor, who "took to his bed" on receipt of the news, and although the company declared that they would not under the circumstances hold him to his agreement, he fell ill, and soon died.

This was the effect, but the discovery of the fault was not the sole cause. Mr. Nowell, the contractor, had been in indifferent health for some time, and the blow coming upon a weakened frame, it gave way. As may be imagined, great excitement and apprehension succeeded the discovery. The directors perceived that their main line was in danger, and that the money already expended on the tunnel was lost.

Some pronounced the tunnel an impossibility; others denounced Robert Stephenson, who they contended should have found out the fault; others, again, wished that some more experienced engineer should be associated with Stephenson, or at any rate consulted on the matter, and the secretary Monsom proceeded to interview Robert Stephenson, their engineer.

He found him busy at work, and intent on the result of the excavations. "What is to be done?" was the first question. "Should not some other advice be summoned?" said Captain Monsom tentatively.

But Stephenson would not hear of that. He considered himself perfectly capable of treeing the tunnel. "I intend to pump out the water," he said. Drain the sand and cut through it was his plan. "Don't be alarmed. Tell the directors that if they will give me time, I will get the water out. If I cannot, then I will ask for assistance; but not till then."

These, or words to the same effect, reassured the secretary, who returned and reported progress to his Board. The directors determined to leave the engineer to carry out his plans. He had shown himself capable hitherto, and they trusted that he would clear the tunnel, though some shook their heads. "The engineer, you see, is such a very young man!"

Stephenson got steam-pumps to work, and tapped the quicksand. About sixteen hundred gallons of water were pumped out of the sand *per minute*; a feat of which those employed were proud, but they did not then anticipate that the pumping would continue for *nine months* without ceasing, and that they had virtually to drain the sand for an extent of several square miles before they could pronounce success assured.

As soon as the necessity for a prolonged spell was recognised, huts were erected on the hill, and Kilsby became an important place. The number of navigators employed exceeded twelve hundred, and their manners and customs did not tend to impress the resident population. We have already touched upon the tastes and habits of the early navvy and his "tally-wife," so we need not again describe them. Suffice it to say that they did pretty much as they pleased, and wrought a good deal of mischief during their stay.

However, if their moral character was indifferent, if a taste

z

for farmers' fowls, and a too pronounced admiration for farmers' daughters were apparent, we must give them credit for pluck and endurance, and almost unlimited powers of work. Take a few cases in connection with this very tunnel, as related by Sir F. Head, and other chroniclers. Sunday did not "crop out" there, they said, and services were neglected; but they certainly did their work on week-days.

On one occasion as the pumping was proceeding, a navvy fell down one of the shafts, and went through the planking into the water in the tunnel with a souse which could be heard all up the shaft. The greatest consternation prevailed; all kinds of handy applications were sought for; shouts and cries went up for assistance, for ropes and pulleys; but when the din was for a moment hushed, a stentorian voice was heard shouting from the bottom of the shaft, " Darn your row; make less noise; coom and help me out!" It is only fitting to state that the injured navvy was rescued, but was compelled to subdue his restless energy in the nearest hospital for the space of some weeks.

Another misfortune befell the sturdy navigators, and it would have had a fatal termination but for the presence of mind and pluck of the ganger in charge. While the water was being pumped out of the tunnel, and while the men were engaged working, the water suddenly sprang out from the roof. The ganger and his men were all bricking the roof, and for that purpose were floating on a raft, for the excavations were all under water. When the quicksand so suddenly burst out in a fresh place the men were not alarmed, but when they perceived that the pumps were unable to restrain the increase of the flood, and that the raft on which they were perched was rising higher and higher, they were in a quandary. If they remained, they would gradually be crushed against the roof of the tunnel. If they swam off, they would be drowned or suffocated; what could they do?

The peril became momentarily greater; there was no exit from the tunnel save by the shaft; how could they reach it?

They had no means of propelling the floating platform, save their working tools, and they could not manage to push by the rugged sides. At length the ganger Lean solved the difficulty. A strong and expert swimmer, he let himself into the water, and one of his gang handed him a rope. With one end attached to the raft, the other between his clenched teeth, the brave leader, assisted by the navvies as much as possible, towed and guided the raft to the bottom of the nearest shaft, where assistance was at hand to haul them all safely to bank.

The influx of water increased so fast that the works had to be suspended, and not all the pumping seemed to make any difference in the tunnel. Despair began to flap her gloomy wings in the faces of the Railway directors. They made up their minds to abandon the work, and cut the line in another direction. Success was "impossible," they said.

When this determination was conveyed to Robert Stephenson he was alarmed. He was as sanguine as ever, but of course could understand the reasons for the decision. Yet he pleaded through his friend, the secretary, for a little respite. "Another fortnight; and then if we can't clear the tunnel, let it alone!"

The reluctant authorities consented, and Stephenson got all his steam-pumps at work. Day by day the water was closely inspected. Hour after hour went by, and the measurements told of the continually flowing tide upon which the pumps had no effect. Some days passed; the soundings showed no improvement. A few days more, and all the efforts of the engineer would come to nought. It was disheartening, vexatious, terrible! One afternoon a gleam of hope penetrated down the working shaft, and a whisper was heard—"The water hasn't gained any!"

A joyful thrill pervaded all those men so deeply interested. The town on the tunnel was disturbed. Was the news true? "They say so. Hurrah!"

There was a general gathering round the shaft. The pumps continued steadily. "Sound again!" The order was obeyed, and the engineer's delight may be imagined when the report came up, "Engines making way; water decreasing!"

Science and Steam had conquered Nature after one of the most stubbornly-contested struggles of our time.

Thenceforward progress was rapid. The twelve hundred and fifty men, the thirteen steam engines, the two hundred horses, united their efforts, and succeeded. The men worked like demons, and little heeded danger or death; they didn't intend to give the water another chance, not they. They worked hard, and when, as occasionally happened, a man fell, or was in some manner killed, his fellows made no pause, they simply pushed the lifeless body aside, and continued the work. But they turned out clean and sober on Sunday to the funeral, and had respect enough for their "mate" to walk bare-headed, and even hand in hand, silently, till the interment was completed, and then they returned to pipe and pot and horse-play. Ere nightfall, perhaps, another adventurous spirit would have gone to its rest—as the most fascinating feat was to leap across the mouths of the shafts, and a failure brought the penalty of death, suddenly, or after lingering agonies. Twenty-six men perished in this and in other more legitimate ways during the construction of the Kilsby Tunnel.

Thirty months, three hundred thousand pounds sterling, and thirty-six millions of bricks were consumed or used in the construction of this tunnel. The reader, when within some six miles of Rugby, on the London side of that town, will plunge into it when next he goes north in the *Flying Scot* or *Wild Irishman,* and will perhaps recall the difficulties surmounted by the engineers of the London and Birmingham (L. & N.W.) Railway.

Then the houses on the hill became untenanted; the navvies, their occupation temporarily gone, dispersed in search of other work. The ankle-jacks have left their thick impressions; the huts have revealed beer barrels—for your navvy is not particular to a licence, he takes it. Comparative silence reigns—no more fighting to decide "who's the best man"; no more squabbles or domestic disputes. The men and their belongings, wives, and children ; the babies and the bundles ; the few articles of domestic use are scattered in search of another " seat of work," where the men will be taken on.

The customary notice to quit is not either given or expected by the contractor. The navvy can dismiss himself at any time. He may leave, go off and work somewhere else, and after a time return to the old job, no questions asked, no reasons given. Independent, hard workers, great eaters, great drinkers (not drunkards), the navvy is nearly always an Englishman. In the days of our railway-making mania nearly all our navigators were Englishmen, natives of England itself, very few Irishmen, who were not tolerated at all. The Scots went in for authority, and were mostly gangers. The pay then was about six shillings a day, now it is about three shillings and sixpence.

It would be easy to write a whole chapter upon navigators, and the temptation to do so is strong upon us ; but we must resist it, and be satisfied with the occasional glimpses which we obtain of this most interesting class in our wanderings in subways.

Numerous other tunnels might be considered, but as the general mode of construction is the same as one or other of those tunnels we have given or shall give as examples, it will not be necessary to do more than mention them. The Box Tunnel, on the Great Western Railway, is one of the best known ; there is a view of the entrance at the head of this chapter. The tunnel is one mile seventy-two chains in length, and is perfectly straight. In spring and autumn the sun shines right into it, at times. This celebrated tunnel occupied two years in construction, and quite one hundred lives were lost in it.

Of Alpine tunnels, the Arlberg and the Brenner could be detailed. The Mersey Tunnel is another fine example of engineering in England. The proposed tunnel under the channel also has claims to some attention ; but these all bear a similarity to the works already described or to be described, as we pass on to the Underground Railways and Electric Lines in the Metropolis of England, concerning which one may collect many interesting and almost forgotten facts.

WAPPING STATION E.L.R.

CHAPTER VII.

THE METROPOLITAN LINES.—OLD PADDINGTON.—THE TRAFFIC
TO THE CITY.—THE WORKING OF THE "UNDERGROUND."
—ITS ROMANCE.

THE Metropolitan Underground Railways are now accepted as such a matter of course, that it hardly seems but "the other day" (comparatively with some of the periods to which we have referred) that Paddington woke up, and was united to London. Indeed, until 1840, when the Great Western Railway invaded the market gardens of the suburb, communication with Paddington was not so easy. A glimpse of the situation in the early years of this century will establish this statement.

From the time of the Saxons until about one hundred years ago, say 1790, the little rustic village of Paddington enjoyed the pleasures of comparative isolation. It is spoken of even less than eighty years ago as a rustic place, about a mile from London, and the road thereto by Oxford Street had a bad name, not only by reason of the ruts, sloughs, and hollows in it, but because of the footpads and highwaymen who˙ infested it.

In the beginning of the century, one Miles ran a pair-horse coach to Paddington, from Holborn and the City; though stage-wagons were run daily in previous years. This Miles' coach was enlivened by the celebrated attendant known as "Miles' boy," who cheered the passengers on the journey to and from the city by playing the fiddle and telling stories to beguile the time wasted in numerous stoppages.

It will readily be understood that some such amusement was necessary, when we mention the astounding fact that more than three hours were consumed in the journey, and for this ride the fares charged were three or two shillings, according to "class"! Paddington Green was a dirty, noisy, neglected place, though in 1801 the Grand Junction Canal to Paddington had opened up communications, and in 1820 the connecting link of the Regent's Canal by Maida Hill Tunnel had put the sometime rustic village in direct touch with Limehouse on the Thames.

Notwithstanding the establishment of "stages," they were not greatly patronized at first. The road had an evil reputation. Books were written exposing the practices of the highwaymen, and suggestions were made that armed parties should escort travellers. But as London grew, "Tyburnia" (where the gallows stood, near the present Connaught Place) became built upon. The London streets became crowded. The railroads gave an impetus to locomotion, and about 1837, several suggestions were made with a view to reducing the congestion

of the thoroughfares. Baynard's watering-place, the meadows
and pools near Paddington, where cattle throve, were invaded.
Baynard's property, or what had been the property of the
former possessor of the "castle" by the river, was gradually
covered with houses, and the district became known as Bays-
water. Residents increased and multiplied, and Mr. Charles
Pearson, the well-known City solicitor, made many practical

PADDINGTON STATION IN 1843.

suggestions, and a tunnel-way around London was proposed,
only to be laughed at. The idea of a railway underground
was derided. All kinds of accidents were prophesied. The
roads would fall in, the passengers would fall out, or be
suffocated ; houses would be undermined, and the occupants
would be killed !

But the idea soon bore fruit. In the year 1834, when

Paddington had increased to a population of about 20,000, and the district of Marylebone numbered nearly 200,000, a railway was proposed and constructed from Paddington, along the Marylebone Road. The steam - engine conveyed City men thus early to Moorgate Street. This line paid, and as it was manifestly impossible to make any more lines on the already too crowded streets, an underground line was projected and derided, as already stated.

The crowding continued and increased, and in 1853, a line from the Edgware Road to Kings Cross was sanctioned. In the following year Parliament agreed to this line being extended, and the "North Metropolitan Railway" to the General Post Office was within "measurable distance" of becoming an accomplished fact. Mr. Pearson had suggested the underground line from King's Cross to Farringdon Street, and the railway companies, already with termini in the outskirts, loudly demanded admittance to London streets.

But the Railway Commission did not see the necessity for all these extensions. It seemed to them the acme of absurdity for any trunk line to want to bring its passengers nearer to business than Waterloo, King's Cross, London Bridge, Paddington, or Euston Square! As to permitting the lines to cross the Thames—absurd! The notion was preposterous! The river navigation must not be impeded by bridges. But the engineers did not lose grip of their ideas. They must go underground; and though the Bill at length passed in 1854, the intelligent public did not see their way to invest money in the undertaking. So the scheme slept until 1860, when "the City," having pronounced in favour of the plan, the required capital was subscribed, and the Metropolitan Railway was born, but not yet made.

However, in 1863, the first lines were completed from Bishop's Road to Farringdon Street, and proved an immense and an immediate success. Crowds awaited the departure of the trains. People who had laughed Mr. Pearson to scorn now came clamouring for tickets; and the Companies' employees could not find room for the would-be passengers. The same scenes were enacted in the evening. The passengers stood patiently or impatiently waiting admission as they do at the pit door of the theatre, and it is wonderful that no very serious accidents were reported.

The 10th of January, 1863, was the red-letter day in Mr. (afterwards Sir John) Fowler's calendar. The railroad underground was a success, after many novel difficulties had been encountered and overcome. Most of us remember the deep excavations in the Euston or Marylebone Road, and other districts, the shoring up of houses, the fears concerning the safety of these dwellings, the pumping, the tunnelling and general excavations.

What did the construction of the underground railways mean? What amount of work do the rival "Metropolitan" and "District" lines represent! The first line was sanctioned in 1854, the "circle" was completed in 1884. So a period of thirty years elapsed before the full benefits of the scheme were reaped by the public. The shareholders, unfortunately, have not reaped corresponding benefits, notwithstanding the enormous number of passengers carried annually.

Each railway company now works its own trains; but the Metropolitan line was originally laid on the broad gauge, though the narrow gauge metals were also put down to Moorgate Street. The Great Western first took the line in hand, but a great deal of heart-burning arose over the engines best suited to the railway.

The hot-water locomotive—an idea of Mr. Fowler's we believe—was discussed, but was not adopted. This invention was designed to obviate the suffocating vapours and the unpleasantness arising from the burning of coal, coke, or other fuel, the heated water preserving a certain quantity of steam for a short journey. This suggestion may have, we do not say it did, influenced the mind of the contractors to build the several stations on the north side almost devoid of means for ventilation, an omission which forced itself disagreeably upon the notice of the travelling public very soon ; and the subject —not the station—was fully ventilated in " the Press."

The hot-water locomotive having been found wanting, a substitute in the fire-brick method was attempted. The plan of this machine was designed to permit the running of the engine by the heat accumulated in the boiler-chamber, and so do without the action of steam in the tunnels. Of course in the cuttings, or places open to the air, the fire-brick engines would be permitted to use steam in the ordinary manner. But the system of accumulated heat did not answer any better than the hot water.

By this time the underground line was approaching completion, and Great Western Railway engines, specially altered, and arranged to suit the novel conditions of traffic, were provided. They ran on the broad seven-feet-gauge, while the Great Northern, not to be behind hand, came down with large, tender-locomotives, which seemed rather out of place in such scenes, and running such short distances.

For seven months the Great Western line kept the traffic going, the Great Northern assisted; but after a while the G.W.R. declined to run the trains at all, and the Metropolitan Railway Company had to find their own power. This they did. Messrs. Beyer & Peacock built the engines—very powerful locomotives they are—and the type has latterly become common. They are constructed so that no smoke interferes with the steam emitted from the small exhausts.

We have no need to detail the work entered upon : the usual incidents of tunnelling are by this time familiar to our readers. But many features of the enterprise were novel. The nature of the soil to be cut through, the various streams and other obstacles to be overcome, the peculiar nature of the tunnels required—one underneath another in places, were almost unique features in the construction of the lines of the Metropolitan and Metropolitan District Railways. To these we may devote a few moments without dealing in technicalities. The various strata through which one or other of the lines run would give food for thought to a student of geology, who would inform us of the remote periods long anterior to the time when Britain was a "tight little island," or indeed an island at all! Even the engineers were unable to fix a date to those deposits.

But they disclose some curious hypotheses. If calculation be correct, the sandy and gravelly soil points to a time before Father Thames was independent. They indicate a period when, if he ran alone he was only a child, a tributary of the grand Rhine, which so proudly walled the great continent, now broken up by the cold North Sea, when Britain was no island, but merely an off-lying portion of the continent; in days when the "silver streak" and the Channel Tunnel—even a tunnel between Ireland and Scotland—were undreamt of; and people could visit France without having any qualms in the matter. A curious age—a curious scene this, which is conjured up by the inspection of some earthy deposits in the cutting of a modern railway! When Britain became an island finally, perhaps, the cave-men disappeared, with other wild animals, from the valley of the Thames, and from the line of the future railway some thirteen feet below its level at high · water.

There were rivers and brooks, now almost forgotten, which intersected the line of utility, and demanded accommodation. The streams have fallen from their original purity and have become sewers. The Tybourne, the Westbourne, the River of Wells, the Fleet, have disappeared in pipes under brick-work as completely as the once pretty hills and woods of Primrose, and Nutting, Snow, or the grove of St. John's have disappeared beneath the avalanche of bricks, stone and mortar. The Westbourne stream, now degraded into Ranelagh sewer, gurgles turbidly along its tube over Sloane Square platforms. The Fleet is met with five times in the twenty miles or so of the Underground Railways. Bridge Creek River, now a sewer, drags itself under Earls' Court Station, and the Tybourne (which gave its name to a district near which rose the gallows) bestowed its title on St. Mary-by-the-Water atte Bourne, and wound down Park Lane, once called by the name of the stream. This Tybourne has become the King's Scholar Pond sewer, and leads a terrible life of it in darkness, amid unhealthy surroundings and occupations.

There are two somewhat remarkable tunnels on the line of the Metropolitan—if it be not superfluous to mention tunnels in the case of an underground railway. There are the "Widening" Tunnel at King's Cross, where the line is double, and the Clerkenwell Tunnel. These are veritable tunnels, as distinguished from excavations. A description of the works and sidings in connection with the Smithfield markets would fill several pages, and, though interesting, would scarcely come within our limits.

As the traffic of the metropolis and of other large cities increases, the necessity for underground or overhead communication—for low or high level lines—has increased. Already the demand for electric railways, in preference to railways worked by steam, is spreading. An underground tramway is already suggested in Paris, worked by electricity from the Bois de Boulogne to the Bois de Vincennes. This tramway will be tubular, of metal, and will be ventilated by electric fans. The mode of construction will be very similar to that employed in the London Subway, which extends from the Monument, in the City, to Stockwell, famed for its "Swan." There were some remarkable features in this undertaking, and as they may be novel to many readers, we will briefly sketch the Electric Railroad—the pioneer line of its kind in this country.

THE PLATFORM AT STOCKWELL STATION.

CHAPTER VIII.

THE LONDON SUBWAY.—A ROMANCE OF THE SEWERS.—SECRET PASSAGES AND CHAMBERS.—CONCLUSION.

"GOING by the sewer, sir?" asked a respectably-dressed man of the writer one afternoon. The individual addressed was not on that occasion intending to burrow mole-like in the earth. He had visited the line before the trains were running. Had he not descended into the clayey shaft mid clang of pumps and roll of skeps? Had he not been personally conducted along a muddy, puddle-y, track, with candle stuck in a lump of clay, to the torture chamber, termed by courtesy an air-lock, where wedges were driven into his ears and head, where cruel compression deprived him of some voice and nearly all hearing, and subsequently caused loss of blood by congestion of the veins? Had he not witnessed, by candle light, the digging out of the soil, where trickling Effra threatened to burst in and sweep away the ghostly, almost silent navvies, as they seemed. Had he not taken his seat in the air-lock full of hope, in possession of all his faculties; and had he not been sent up to bank muddy, dishevelled, half dazed, more than half deaf, bleeding from the nose and generally compressed?

Yes; all these things had happened; so the writer was not anxious to return through the Sewer, even though it was now swept and garnished, fitted with trams, and perfectly free from all annoyances. No smoke, no vapour, no oppressiveness. The writer's Romance of the Subway may be briefly related.

A gang of men going out with the light; a gang of men going into the night. The off-shift and the on-shift. The visitor accompanied a few of the latter, who made no objection and no remark as the whole of the party trudged manfully through the subway, girded up with iron rings bolted together, and the spaces of any filled in, when found, with cement.

In front the men ranged themselves along the right-hand side of the subway. An iron door stood closely shut beyond a trolly-rail, cut through the " room " in which the men were seated; and no sooner had the writer and his conductor entered it, than another air-tight door was shut, and we all sat in dead silence within the sealed and secret chamber.

Here was a study for the visitor—a romance in itself. Shut up with the men, all confederates, what dire deed might not be committed? Sworn to secrecy, the gang could rob and dispose of the body of any rich and " paying " visitor. Was it imagination? Were the men regarding the stranger with curiosity and suspicion? They were, undoubtedly. Could he not get out?... No... A hissing sound succeeded the turning of a tap. ".Hold up," whispered some one. Still the men stared. What had happened? Why could no voices be heard? Why were they all dumb? Why did ears and temples throb to bursting, almost? Why was an iron band screwed down upon the forehead, and wedges painfully inserted into the ears, within which blood surged and hummed, making the only audible sounds.

The further iron door was then opened; a little of the pressure was soon removed. The hearing gradually returned to some extent, but talking was apparently in whispers. In front were men digging by means of a shield, pressed by hydraulic rams, which cut away the earth in huge slices that fall within reach of the navvies, who clear the soil out through a cavity in the centre of the shield. The use of compressed air was necessary just there, because the water could not be pumped out of the pulpy sand, where Effra still holds sway, or recently held sway; that same river upon whose floods did the Virgin Queen travel in her state barge to Raleigh House, in Brixton. A curious reminiscence, this, to come down from Tudor days when the country round was wild, and open to the air, to the Victorian era of vistas of houses and compressed atmosphere under ground! Yet the Effra carried it all these years, and yielded up the memory to the visitor in the compressed air chamber under the roadway.

The construction of this subway was completed without any disturbance of the roads, houses, pipes, or sewers. Side by side, or one above the other, according to the exigencies of space, the two lines are laid. The upper and lower lines answer to "up" and "down" lines most completely. Thus the two lines are quite separate in fact, each in a tunnel of its own. The manner in which this subway or subways pushed along, amid pipes and channels of gas and water, worm-like avoiding obstacles and without trespassing on "vested rights," is astonishing.

The descent is some sixty feet, but the lifts, as in the Mersey Tunnel and other undertakings, provide for this. The pushing shield, the hydraulic pressure, the flow of water, and compressed air, aided some 4,000 men to cut their way, without causing any disturbance or inconvenience to any one, for more than three miles under our most frequented suburban roads.

The South London Subway is unique of its kind at present,

but other electric railways are projected, and the novelty will soon wear off. Nevertheless, it has a future before it. The electric railway will be the "boom" of the "coming race." . . .

London is honeycombed with subways, and, of course, other cities, in civilized countries, are equally perforated beneath the streets. The Holborn Viaduct and Thames Embankment supply specimens of the varied tunnels "of sorts" which have been driven beneath our roadways. Follow that man who is disappearing, like a demon down a stage trap, into a square opening in the roadway or pavement. There is one within a hundred yards of the writer, down which jack-booted, fearsome-looking men disappear, or from which they unexpectedly and cautiously emerge. If those who go down to the sea in ships see wonders, what do not those who go down into the sewers witness?

Can you picture the man or men by lantern light pursuing their slimy way along the tubes which ramify beneath our thoroughfares, amid the noisome rats and not too wholesome atmosphere. How many men have been lost, think you, in these queer passages? How many may have gone down in companionship, and have returned alone? What was that terrible story of a man who was so jealous of a sailor who lodged with him that he persuaded him to accompany him on his subway rounds, and the other man, innocent and unsuspecting, consented, believing that the elder was afraid, and growing timorous?

How did it happen one day that the elder man came up alone, and returned home without saying anything of his companion, that his young wife made inquiry, but was put off with evasive answers, and with so many quibbles, that her suspicions were aroused. She made further search, and discovered that the young man had never been seen since he had descended into the great sewer, the outfall sewer of the town, with her husband.

She was a loyal wife, and though she feared the worst, she

said nothing to incriminate her husband, but lived on. Months
passed, years passed ; nothing was heard of the missing man,
and the husband had relinquished sewer-inspection for the
more lucrative employment of keeping a public-house. One
night a party of sailors came in, and one of them made inquiry concerning a friend who had been lodging with the innkeeper in old days when he was "sewer-man."

The landlord grew nervous, but gave evasive replies ; the
sailor pressed his questions, so did his messmates. They
intended to make inquiry, and find out whether the man had
been murdered ! They would have the drain searched. So
they left again after some conversation with the landlady, who
was most specially interested ; even going so far as to say that
she could perhaps give them a clue.

The landlord quaked. He shut up the house in terror of
his wife, who would, he believed, betray him. Next morning
the house remained shut—it was broken open ; the landlord
and his wife were found, both dead ; and when the sailor
returned in pursuance of his promise, he was shocked to find
that his joke had had such a ghastly termination.

For he was the lately missing man ; the husband had left
him in the sewer by accident, a storm came on, then the water
rose and poured out—the husband did not warn the younger
man, and when he himself returned, he felt that accident had
carried his rival, as he thought him, away. But his wife's
questions and doubts alarmed him ; he became frightened.
Had he really murdered the man ? His wife did not credit his
version of the affair, and he could not bear to visit the outfall
again—hence his misery.

Meantime the young man had escaped, and judging from
some previously plain spoken words of his landlord of the state
of his mind, he quietly disappeared and went to sea. He and
his friends had arranged a practical joke, and took the landlady
into their confidence. She, delighted at her lodger's return,
and her husband's innocence, assented to the plan to surprise
him ; but his conscience proved too strong, his fears conquered,
and he killed his wife and himself to avoid detection.

Many tales of the underground passages, the catacombs and other subterranean ways, could be told almost *ad nauseam.* But space will not permit of further narrative. We are compelled to pass very lightly the records of the "engineering" of old manor houses and halls. Take almost any mansion, and you will find the secret panel opening into the secret passage, which conducts the searcher to the hiding-place of the fugitive. Carew Castle can show such passages cut in the thickness of the walls. Whiteladies, Hindlip Hall, and numerous others possess such "tunnels," in which priest and prince have at one time or another taken refuge, or by means of which they have escaped.

Every castle used to have its underground passage, and the same means of communication with the outer world were to be found between monasteries and the cathedrals hard by; for instance, between Durham Cathedral and the Abbey of Finchale there is such a passage. Another leads under the Severn, from Bridgnorth; and a third, discovered by the sinking of the road, exists by Buildwas Abbey, on the same classic river, and so on, almost *ad infinitum.*

The antiquarian could inform us, if he would, of many another underground passage. He could lead us to Broughton Castle where Lord Saye and Sele's hidden and secret conclave met, whose members assembled at a distance from the castle, and then by the "underground" reached the council room. We need not go to Woodstock, for our illustrations of the tricks played, the "ghosts" seen, and the manifold marvels which were developed in the Middle Ages by means of hidden traps, and concealed or underground passages. Boscobel, Sutton Place, Calverley Hall, Moseley Hall, Lydiate House or Hall, and many other interesting mansions, can supply us with legendary and truthful records sufficient to make a separate section of "underground romance," with a ghostly flavour, which is, however, unsuited to such a "matter of fact romance" as this!

So we pass on, for we must now conclude these chapters in

the romantic side of Engineering, not from any failure of material, but because the volume may otherwise become unwieldy.

Material, indeed ! What anecdotes and records might not be related and re-written of the Water Way and the Air Way, the Smuggler, the Lightship, the Aeronaut, the Telegraph? A volume in itself could be indited of Railway Romance, of which the history, in the words of our latest Railroad historian, "yet remains to be written." And so it is. The experience of one accustomed to the railroad, in practice, is something very different from that enjoyed by the mere traveller or compiler of records.

Yet it is with real regret that we find ourselves at the end of a volume which has proved almost a labour of love. The incidents vary in interest, it may be ; they are not detailed for students of old standing, or for engineers, to whom such events are so familiar, but for the thousands of youths and maidens and non-technical people who would like to trace the marvels of our modern engineering back to their initiation and modest beginnings. If such readers are pleased then the author will be more regretful still that the time has come for him to write

THE END.

INDEX.

357

St. Etienne, burning coal in, 279.
St. Gothard, grand road, 56, 303;
Hospice of, 307.
St. Mary-by-the-water atte Bourne,
348.
St. Paul's Churchyard, 4.
Stephenson, George, and the Man-
chester and Liverpool Directors,
190, 191; his engine the
"Rocket," 192; and Railway
Clearing House, 239; and
"narrow gauge," 259; men-
tioned, 186, 279, 281.
Stephenson, Robert, his opinion of
Wear Bridge, 38; and first use
of iron railroad, 179; and the
amateur tunnel maker, 316; men-
tioned, 195, 198, 200, 335, 339.
Stockton and Darlington Line, 186,
188, 190.
Straining Tower, Vyrnwy Lake,
173, 175.
Stratæ Viæ, 3.
Strike of workmen at St. Gothard,
309.
Subways, 265, 352.
Sudbroke, Roman Encampment at,
318, 319.
Sudbrook Village, 320; works
"drowned," 327, 330.
Sunday Travelling, opinions on, 205.
Sutton Place, 355.
Suwarow, fatal retreat of, 58.
Swallow, the (engine), 260.
St. Michel, 290, 291.
Saltash Bridge, 260.
Samos, tunnel at, 268.
Sandars, Mr., 185, 194.
Sankey Brook, 120.
Sanspareil, the (engine), 192.
Sardinia, king of, 291.
Savery, 2, 272.
Savoy, 291, 294.
Saxby, Mr., 228.
Saye and Sele, Lord, 355.
Schuylkill, River, 37.
Sedan Chairs, Mr. Sala's definition
of, 61.
Semaphore, the, 228.
Semmering Pass, the, 293; railway,
307.
Sempronius, old path of, 58.

Severn, the, 38, 319, 321, 325, 327,
355.
Sewers, terrible story of, 353.
Sharpe, 194.
Sheffield Colliery, 179.
Shrewsbury Castle, 34; fall of
church tower at, 35.
Siberia, mining in, 268.
Sibthorpe, Col.; objection to rail-
roads, 200, 203.
Signalmen, 232.
Signals, railway, 222, 227, 230.
Simonia, M., pathetic incident re-
lated by, 269.
Simplon, the, 2, 58, 290, 301.
Sismonda, Signor, 291.
"Sixteen String Jack," 18.
Skip Bridge, 25.
Slidell and Mason, incident, 216.
Smeaton, John, 21, 158.
Smiles, Mr., his history of the con-
quest of Chat Moss, 189; quoted,
32, 33, 34, 65, 116.
Soames, Mr., and London Water
Supply, 158.
Society of Friends, association of,
185.
Sommelier, Signor, 293, 303.
Sostegni barricade, 119, 120.
Southbrooke, John, 318.
South Wales, change of line in,
262.
Spedding, of Whitehaven, 280.
Speed of rival lines, 263.
Splugen, the, 57, 58.
Sweden, mines in, 269.
"Switch," invention of, 228.

"Tale of Two Cities," description
of Coaches in, 59.
Taff, River, the, 22.
Tamar, River, the, 282.
Tees, River, 271.
Telford, John, 32.
Telford, Thomas, public acts of,
38; and Macadam's method, 42;
Rickman's Life of, 44; and Sus-
pension bridges, 49; Southey's
lines on, 52.
Thames Police, the, 109.
River, 111, 155, 348.
Tunnel, 260.

Warwick House
Salisbury Square
LONDON E.C

A LIST OF

New and Popular Books

PUBLISHED BY

WARD LOCK & BOWDEN LTD

GUY BOOTHBY

In Strange Company. A Story of Chili
and the Southern Seas. By GUY BOOTHBY, Author of
"On the Wallaby." **With Six Full-page Illustrations**
by STANLEY L. WOOD. Crown 8vo, cloth gilt, bevelled
boards, **5s.**

Mr. Boothby's new book fully justifies its title. It is the
story of remarkable adventures encountered in strange company, and it is told with sufficient power, picturesqueness,
and originality, to completely fascinate the reader.

OUTRAM TRISTRAM

The Dead Gallant; together with "*The
King of Hearts.*" By OUTRAM TRISTRAM. **With
Full-page Illustrations** by HUGH THOMSON and
ST. GEORGE HARE. Crown 8vo, Irish linen gilt, **5s.**

No plea need be put forward now for the Historical
Romance, and Mr. Outram Tristram's success in making
days that are gone to dawn again before the reader is very
striking. He is no unworthy compeer of Mr. Stanley Wey-
man and Mr. Doyle, and these two romances—the first of
which deals with the famous Babington Conspiracy, and the
second with the Young Pretender—are triumphs of the
novelists' art.

A. CONAN DOYLE

A Study in Scarlet. By CONAN DOYLE, Author of "Micah Clark," "The Sign of Four," "The White Company," etc. Fifth Edition. Crown 8vo, cloth gilt, **3s. 6d. With Forty Illustrations** by GEORGE HUTCHINSON.

"Mr. Conan Doyle's stirring story of love and revenge well deserves the honour of a third edition."—*Saturday Review.*

"It is very good. . . . Sensational, crisply written, and exciting."—*Review of Reviews.*

"Few things have been so good of late as Mr. Conan Doyle's 'Study in Scarlet.'"—Mr. ANDREW LANG, in *Longman's Magazine.*

"One of the cleverest and best detective stories we have yet seen. . . . Mr. Conan Doyle is a literary artist, and this is a good specimen of his skill."—*London Quarterly Review.*

J. E. MUDDOCK

Stormlight; or, the Nihilist's Doom. A Story of Switzerland and Russia. By J. E. MUDDOCK, F.R.G.S., Author of "For God and the Czar," etc. Eighth Edition. **With Two Full-page Illustrations** by GORDON BROWNE. Crown 8vo, cloth gilt, **3s. 6d.**

"Interest is sustained all through in so exciting a plot, and the story may be recommended."—*Lloyd's News.*

"Strong in dramatic incident, and highly sensational; the reader's interest never flags for a moment."—*Manchester Guardian.*

"The work has a strong plot, exciting situations, and a certain truth to history that make it full of interest."—*The Scotsman.*

WARD LOCK & BOWDEN LTD

JOSEPH HOCKING

Ishmael Pengelly: an Outcast. By

JOSEPH HOCKING, Author of "The Story of Andrew Fairfax," etc. With **Frontispiece and Vignette** by WALTER S. STACEY. Crown 8vo, cloth, **3s. 6d.**

"The book is to be recommended for the dramatic effectiveness of some of the scenes. The wild, half-mad woman who has been wronged is always picturesque whenever she appears, and the rare self-repression of her son because of the girl whom he loves, is admirably done."—*The Athenæum.*

"The critical point in the book is finely managed, and the whole story is told with quite unusual power and a large measure of trained skill. Mr. Hocking has produced a novel which may unhesitatingly be recommended to all classes of readers."—*The British Weekly.*

The Story of Andrew Fairfax. By

JOSEPH HOCKING, Author of "Ishmael Pengelly," etc. With **Frontispiece and Vignette** by GEO. HUTCHINSON. Crown 8vo, cloth gilt, **3s. 6d.**

"The really excellent part of the book is its accurate picture of the monotony of rural life. . . . A readable, wholesome, and carefully written story." — *Westminster Gazette.*

"The author writes of a country folk whose troubles and needs he knows, and whose prejudices he can see, and without losing his hope for them. He is full of quiet humour, too, and the book is pleasant reading throughout."—*Literary World.*

The Monk of Mar Saba. By JOSEPH

HOCKING, Author of "Ishmael Pengelly," etc. With **Frontispiece and Vignette** by WALTER S. STACEY. Crown 8vo, cloth gilt, **3s. 6d.**

"Of great power and enthralling interest. . . . The scenery of the Holy Land has rarely been so vividly described as in this charming book of Mr. Hocking's."—*The Star.*

WARD LOCK & BOWDEN LTD

HENRY FRITH

The Romance of Navigation and

Maritime Discovery. From the Earliest Periods to the 18th Century. **With about 120 Engravings.** Crown 8vo, cloth extra, full gilt, **3s. 6d.**

"A capital boy's book. . . . Bright, entertaining, and satisfactory. . . . . Handsomely got up, admirably printed, and enriched with a multitude of illustrations."— *Aberdeen Free Press.*

The Romance of Engineering : Our

Highways, Subways, Railways and Waterways. **With 150 Illustrations.** Crown 8vo, cloth gilt, **3s. 6d.**

The *Daily Telegraph* says—"Those who desire to combine entertainment with amusement could not do better than present an intelligent youth with a copy of the 'Romance of Engineering.' "

ETHEL S. TURNER

Seven Little Australians. With

Twenty-six Illustrations by A. J. JOHNSON. Crown 8vo, cloth gilt, **3s. 6d.**

The work has all the simple domestic interest of Miss Alcott's "Little Women," with an added lightness and delicacy of touch recalling "Little Lord Fauntleroy," as well as all the delightful fun and humour which made the success of "Helen's Babies." The book will appeal to readers of every class and every age. Of all the children, "Judy," the lovable little tomboy and madcap, is likely to become as famous a figure in fiction as "Topsy" of "Uncle Tom's Cabin."

REV. JOHN J. POOL

The Land of Idols ; or, Talks with Young

Folks about India. With about **120 Illustrations.** Crown 8vo, cloth gilt, **3s. 6d.**

"The young person of either sex who failed to be fascinated by this book would be a very extraordinary character." —*Liverpool Mercury.*

WARD LOCK & BOWDEN LTD

ARTHUR LEE KNIGHT

The Mids of the "Rattlesnake"; or,
Thrilling Adventures with Illanum Pirates. **With Illustrations** by W. S. STACEY. Crown 8vo, cloth gilt, 2s. 6d.

" A stirring sea story, with plenty of fun and adventure to satisfy the most voracious reader. The loss of the *Rattlesnake* and her subsequent recapture, with plenty of pirates and Malays, make up a regular boys' book. The pictures are good."—*The Sheffield Independent.*

The Rajah of Monkey Island. With
Illustrations by W. S. STACEY. Crown 8vo, cloth gilt, 2s. 6d.

"There is plenty of dash and spirit in 'The Rajah of Monkey Island,' and the writer may be quite satisfied that no boy will take up the book without finishing it with breathless interest. . . . All lovers of sailors and the sea will appreciate this excellent yarn."—*Daily Telegraph.*

A Sequel to the Above

The Cruise of the "Cormorant." With
Illustrations by W. S. STACEY. Crown 8vo, cloth gilt, 2s. 6d.

"A rousing tale of adventure by Mr. A. Lee Knight, whose talent for work of this kind is so well known, and so highly appreciated. . . . Full of sensation and excitement, and spiritedly illustrated."—*Glasgow Herald.*

Dicky Beaumont; His Perils and Adven-
tures. **With Illustrations** by W. S. STACEY. Crown 8vo, cloth gilt, 2s. 6d.

"Thoroughly entertaining. . . . This is exactly the book for spirited lads with a taste for salt-water life."—*Yorkshire Post.*

WARD LOCK & BOWDEN LTD

B B

JOHN C. HUTCHESON

The Black Man's Ghost. A Story of

the Buccaneer's Buried Treasure or the Galapagos Islands. By JOHN C. HUTCHESON. **With Full-page Illustrations** by W. S. STACEY. Crown 8vo, cloth gilt, **2s. 6d.**

" This is an exciting tale of adventure and of buried treasure . . . told with spirit, and admirably illustrated."—*Glasgow Herald.*

FRANKLIN FOX

Frank Allreddy's Fortune ; or, Life on

the Indus. The Story of a Boy's Escape from Shipwreck, his Perils, and Adventures in India. By Captain FRANKLIN FOX, Author of " How to Send a Boy to Sea," etc. **With Full-page Illustrations** by W. S STACEY. Crown 8vo, cloth gilt, **2s. 6d.**

" A rattling story of life at sea and in India, which we can cordially recommend. . . . Teems with exciting interest and hairbreadth escapes."—*Review of Reviews.*

R. M. FREEMAN

The Heir of Langridge Towers ; or, The

Strange Adventures of Charlie Percival. By R. M. FREEMAN. **With Illustrations** by W. S. STACEY. Crown 8vo, cloth gilt, **2s. 6d.**

" Handsomely bound and cleverly illustrated, this book will prove a most acceptable present for young lads fond of fun and fighting. . . . The scenes and incidents, exciting adventures and rollicking humour, make up a first-class story."—*Newcastle Chronicle.*

WARD LOCK & BOWDEN LTD

MARY E. WILKINS
A Humble Romance, and other Stories.

By MARY E. WILKINS, Author of "A New England Nun," etc. **With Frontispiece and Vignette** by GEO. HUTCHINSON. Crown 8vo, cloth gilt, **2s. 6d.**

Miss Wilkins has taken a place quite in the front rank of writers of short stories. These charming stories are without an equal in their way.

By the same Author.—A Book for Children.
The Pot of Gold, and other Stories.
With Illustrations. Crown 8vo, cloth gilt, **2s. 6d.**

"Every one who has read the simple little stories of New England life, which Miss Wilkins tells with so much skill, will have perfect confidence in her power to interest children. Of that power she gives fullest evidence in 'The Pot of Gold.' She can tell the oddest little romances in the gravest fashion, always writing understanded of young people. This is a book they will like."—*Yorkshire Post.*

CATHERINE J. HAMILTON
Women Writers: their Works and Ways.

First Series. Including Fanny Burney, Madame de Staël, Jane Austen, Maria Edgeworth, etc. Crown 8vo, cloth gilt, **2s. 6d.**

"Entirely delightful. For a young girl with bookish tastes it will make an ideal present."—*Review of Reviews.*

Women Writers: their Works and Ways.

Second Series. Including Mrs. Hemans, Harriet Martineau, Letitia E. Landon, Mrs. Browning, Charlotte Brontë, Mrs. Gaskell, "George Eliot," etc. **With Portraits.** Crown 8vo, cloth gilt, **2s. 6d.**

"We do not remember having often seen this sort of work so pithily and pleasantly done."—*Literary World.*

WARD LOCK & BOWDEN LTD

SARAH TYTLER
Author of "Citoyenne Jacqueline," etc.

1. *Days of Yore.*
2. *A Hero of a Hundred Fights.*
3. *Papers for Thoughtful Girls.*
4. *The Diamond Rose.*
5. *Heroines in Obscurity.*
6. *Girlhood and Womanhood :* The Story of some Fortunes and Misfortunes.

Each with Frontispiece and Vignette by WALTER S. STACEY, Crown 8vo, handsomely bound, cloth gilt, **2s. 6d.** each.

" We have over and over again heard parents speak something in this style : ' We are at no loss for books for our boys ; there are Mr. Smiles' volumes and others ; but where to look for a good girl's book, a good companion to a young lady just leaving school, we know not, and would be glad for any one to help us.' This complaint need no longer be heard. Miss Tytler's books are exactly of the kind desiderated, and may with all confidence be recommended at once for their lofty moral tone and their real artistic qualities, which combine to make them equally interesting and attractive."—*Nonconformist.*

MRS. WHITNEY

Ascutney Street : A Neighbourhood Story. **With Illustrations.** Crown 8vo, cloth gilt, **2s. 6d.**

" The story is told in a charming fashion, and its moral is one that needs enforcement in our day."—*Literary World.*

A Golden Gossip : Neighbourhood Story Number Two. **With Frontispiece and Vignette** by GEO. HUTCHINSON. Crown 8vo, cloth gilt, **2s. 6d.**

" The character sketches contained in it are smart and full of individuality. . . . The portrayal of the *beautiful character* of the ' golden gossip ' herself is exceedingly clever."—*Nottingham Guardian.*

WARD LOCK & BOWDEN LTD